TM 3-23.25(FM

SHOULDER-LAUNCHED MUNITIONS

Maneuver Center of Excellence
U.S. Army Training and Doctrine Command
Headquarters, U.S. Department of the Army, September 2010

Published by Books Express Publishing
Copyright © Books Express, 2010
ISBN 978-1-907521-33-1
To purchase copies at discounted prices please contact
info@books-express.com

*TM 3-23.25 (FM 3-23.25)

Technical Manual
No. 3-23.25

Headquarters
Department of the Army
Washington, DC, 15 September 2010

Shoulder-Launched Munitions

Contents

Figures

Tables

Preface

This technical manual (TM) provides technical information and training and combat techniques for shoulder-launched munitions. Intended users include leaders and designated Soldiers who will use this information to successfully integrate shoulder-launched munitions into combat operations. This TM also discusses training for proficiency with shoulder-launched munitions.

This TM revision includes references to new materiel and systems and includes—

- The new training strategy to include specific strategies for the USAR and the ARNG.
- New observation devices were added, and outdated ones were removed.
- The M136A1 AT4 confined space (AT4CS) was added, along with its associated training.
- New training aids, devices, simulators, and simulations (TADSS) were added.

This TM prescribes—

- DA Form 7676 (Day and Night Fire—M141 BDM [BDM Subcaliber Training Launcher]).
- DA Form 7677 (Day and Night Fire—M136 AT4 [M287 Subcaliber Training Launcher]).
- DA Form 7678 (Day and Night Fire--M72 [M72AS 21-mm Subcaliber Training Launcher]).

This TM applies to the Active Army, the Army National Guard (ARNG)/Army National Guard of the United States (ARNGUS), and the United States Army Reserve (USAR).

The proponent for this TM is the U.S. Army Training and Doctrine Command (TRADOC). The preparing agency is the Maneuver Center of Excellence (MCoE). You may send comments and recommendations by any means (U.S. mail, e-mail, fax, or telephone) as long as you use DA Form 2028 (Recommended Changes to Publications and Blank Forms) or follow its format. Point of contact information is as follows:

E-mail:	benn.29IN.229-S3-DOC-LIT@conus.army.mil
Phone:	Commercial: 706-545-8623
	DSN: 835-8623
Fax:	Commercial: 706-545-8600
	DSN: 835-8600
U.S. Mail:	Commander, MCoE
	ATTN: ATSH-INB
	6650 Wilkin Drive, Building 74, Room 102
	Fort Benning, GA 31905-5593

Chapter 1

INTRODUCTION

Currently, no single disposable shoulder-launched munition with a multi-target capability exists. This publication addresses the capability gap with three separate types of munitions (Figure 1-1):

- M72-series shoulder-launched munitions, which include—
  - M72A2/A3.
  - M72A4/5/6/7 (improved M72).
- M136-series shoulder-launched munitions, which include—
  - M136 AT4.
  - M136A1 AT4 confined space (AT4CS).
- M141 bunker defeat munitions (BDMs).

The unit issues these launchers as rounds of ammunition to individual Soldiers.

USES

1-1. The purpose of shoulder-launched munitions is to provide the Soldier with a lightweight, disposable, man-portable, self-contained, one-shot system that is highly effective in incapacitating personnel located within protective barriers, such as buildings, fighting positions (earth and timber bunkers), light-armored vehicles, and other field fortifications.

1-2. While all shoulder-launched munitions are useful in damaging or destroying targets, their difference lies in the types of targets they are used against:

- The M141 BDM is designed to—
  - Destroy earth and timber bunkers.
  - Breach 8-inch reinforced concrete walls and 12-inch triple-brick walls.
  - Destroy or collapse underground openings.
  - Destroy unarmored (cars and trucks) and light-armored vehicles, but has very little effect on heavy-armored vehicles.
- M72- and the M136-series shoulder-launched munitions are designed to—
  - Penetrate and destroy light-armored vehicles.
  - Damage older model battle tanks (when fired in pairs or in volley).
  - Neutralize fortified firing positions, and personnel and weapons behind barriers.

DESCRIPTION AND CAPABILITIES

1-3. Table 1-1 provides a description of the types of shoulder-launched munitions and allows for a comparison of capabilities.

COMPONENTS

1-4. Shoulder-launched munitions consist of an unguided free-flight rocket and a launcher that contains all features and controls necessary to aim, fire, and engage targets.

LAUNCHER

1-5. The launcher has a design similar to the recoilless rifle. It is man-portable and provides water-resistant protection for the rocket during storage, transportation, and use. All propulsion unit operation occurs within the launch tube. Table 1-2 depicts the difference between the launchers of various shoulder-launched munitions.

Note. The weights and measurements are approximate. See the corresponding technical manual (TM) for more information.

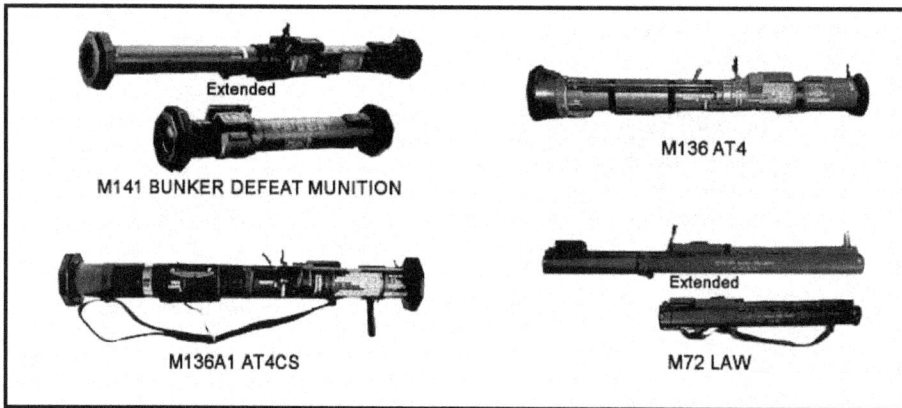

Figure 1-1. Current shoulder-launched munitions.

Table 1-1. Description and capabilities of shoulder-launched munitions.

TYPE	DESCRIPTION AND CAPABILITIES	
M141 BDM	The M141 BDM consists of a free-flight, fin-stabilized, multipurpose munition that consists of an 83-mm high-explosive (HE), dual-mode rocket sealed in an expendable launcher that also serves as a watertight transport and storage container. The M141 BDM addresses the need to destroy hardened targets, such as bunkers and other fixed enemy positions, and incapacitate the enemy personnel inside. It can be employed effectively against double-reinforced concrete walls up to 8 inches thick, triple-brick structures, and standard earth and timber bunkers. It can also perforate up to 20 mm of rolled homogenous steel, which provides a capability against light-armored and thin-skinned vehicles. The M141 BDM can be employed in limited visibility, with the aid of night vision devices (NVDs) or with artificial illumination.	
M136-Series Shoulder-Launched Munitions	The M136-series shoulder-launched munition consists of a free-flight, fin-stabilized, antiarmor rocket packed in an expendable launcher that also serves as a watertight transport and storage container. M136-series shoulder-launched munitions can be employed in limited visibility, with the aid of NVDs or with artificial illumination.	
	M136 AT4	The M136 AT4 is primarily designed for use against armored vehicles.
	M136A1 AT4CS	The M136A1 AT4CS is similar to the M136 AT4, but uses a different propulsion system. This system enables the M136A1 AT4CS to be fired from an enclosure, a capability that the M136 AT4 does not have.
M72-Series Shoulder-Launched Munitions	The M72-series shoulder-launched munition consists of a free-flight, fin-stabilized, antiarmor rocket packed in an expendable launcher that also serves as a watertight transport and storage container. Available M72 launchers are divided into two groups: the M72A2/A3 and the M72A4/5/6/7 (improved M72). M72-series shoulder-launched munitions can be employed in limited visibility, with the aid of NVDs or with artificial illumination.	
	M72A2/A3	The M72A2/A3 is a first-generation munition. It was primarily designed for use against the older generation of armored tanks and light-armored vehicles.
	M72A4/5/6/7 (Improved M72)	The improved M72 warhead differs from the M72A2/A3 warhead in that the improved M72 contains small fin blades and a small copper cone, as opposed to the folding fins and larger copper cone on the M72A2/A3 warhead. The improved M72 is primarily designed for use against the improved armor of light-armored vehicles.

Table 1-2. Launcher specifications.

		M141 BDM	
LENGTH	Extended/Ready To Fire	55 inches	
	Closed/Carry	32 inches	
WEIGHT (TOTAL)	16 pounds		
COMPONENTS	Launch Tubes	Inner and outer filament-wound composite tubes, which are stored one within the other to shorten the carry length	
	Firing Mechanism	Electrical, arms munition when the firing mechanism cover is fully opened *Note.* The safety and trigger buttons are exposed only when the cover is opened.	
	Sights	Front	Rifle-type, three posts (a central post for engaging stationary or moving targets head-on or straight away, and side posts for engaging targets moving left or right)
		Rear	Sight blade, range adjustment knob, range scale, 2-mm peephole for normal daylight visibility conditions, and 7-mm peephole for limited visibility conditions; 500-meter range indicator, graduated in 50-meter increments
		Night	Can be fitted with the AN/PAQ-4C, AN/PEQ-2, AN/PVS-4, or AN/PAS-13 using a fixed mounting rail
	Shoulder Stop	Folding metal and rubber bracket designed to rest against the firer's shoulder to support the launcher while aiming and firing	
	Sling	Adjustable for carrying the munition and providing firing support	
	Safeties	Transport Safety Pin	Secures the munition in the closed position *Note.* The launcher will not arm if the tube is not fully extended and locked.
		Firing Mechanism Cover	Protects the firing mechanism and arms the munition when fully opened (exposes the safety button and trigger button)
		Safety Button	Unlatches the trigger button when pressed and held
		Note. The launcher will not arm if the firing mechanism cover is not all the way forward against the tube and the safety button is pressed.	
	Trigger Assembly	To initiate launch, the firer must press a red trigger button (exposed when the firing mechanism cover is opened).	
	Front and Rear Bumpers	Designed to absorb the shock of daily handling and transport	
		Front	Has an environmental fire-through muzzle cover
		Rear	Has an inner environmental seal (blown away during launch)
COLOR/ MARKINGS	Olive drab green with yellow color-coded band (HE warhead), gold color-coded band (field handling trainer [FHT]), or no band (BDM 21-mm subcaliber trainer) located on the forward end of the launcher BACK BLAST DANGER AREA — FWD ROTATE TO LOCK OPERATING INSTRUCTIONS DECAL SAFE ON OUTSIDE OF COVER ARMED THE ARMED DECAL IS LOCATED UNDER THE FIRING MECHANISM COVER AND IS SEEN ONLY WHEN THE FIRING MECHANISM COVER HAS BEEN OPENED. FWD — TARGET COVER OPENED ARMED DECAL THE SAFE DECAL IS ON TOP OF FIRING MECHANISM COVER		

Table 1-2. Launcher specifications (continued).

M136 AT4			
LENGTH	40 inches		
WEIGHT (TOTAL)	15 pounds		
COMPONENTS	Launch Tube	Single fiberglass reinforced launch tube	
	Firing Mechanism	Mechanical; consists of a red trigger button, an enclosed firing rod and spring, and three safety devices	
	Sights	Resemble those of the M16-series rifle—	
		Front	Sight blade with a center post and left and right lead posts
		Rear	Sight blade, range adjustment knob, range scale, 2-mm peephole for normal daylight visibility conditions, and 7-mm peephole for limited visibility conditions; 500-meter range indicator, graduated in 50-meter increments
		Night	Can be fitted with the AN/PAQ-4C, AN/PEQ-2, AN/PVS-4, or AN/PAS-13 using the nightsight mounting bracket (NSN 5340-01-391-3004)
	Shoulder Stop	Folding metal and plastic bracket designed to rest on the firer's shoulder to support the launcher while aiming and firing **Note.** When not in use, it is snapped to the underside of the launcher.	
	Sling	Adjustable for carrying the munition and providing firing support	
	Safeties	**Note.** The munition cannot be fired until all three safeties have been disengaged.	
		Transport Safety Pin	Blocks the firing pin from striking the cartridge percussion cap
		Cocking Lever	Has two positions: SAFE and cocked • When the munition is in the SAFE position, there is no engagement between the firing rod and the trigger. • When the munition is cocked, the firing rod is engaged with the trigger.
		Red Safety Release Catch	Connected to the firing rod, prevents the firing rod from striking the firing pin when disengaged **Note.** To disengage the red safety release catch, press it down and hold.
	Trigger Assembly	To initiate launch, the firer must press the red trigger button located on the right side of the launcher in tandem with the red safety release catch.	
	Front and Rear Bumpers	Designed to absorb the shock of daily handling and transport	
		Front	Has an environmental fire-through muzzle cover
		Rear	Encloses an inner venturi and baseplate (blown away during launch)
COLOR/ MARKINGS	Olive drab green with black-yellow-black color-coded band (HE antiarmor round, after 1998), black color-coded band (HE antiarmor round, prior to 1998), gold color-coded band (FHT), or no band (M287 9-mm tracer bullet trainer) located on the forward end of the launcher 		

Table 1-2. Launcher specifications (continued).

M136A1 AT4CS			
LENGTH	41 inches		
WEIGHT (TOTAL)	17 pounds		
COMPONENTS	Launch Tube	Single fiberglass reinforced launch tube	
	Firing Mechanism	Mechanical; consists of a red trigger button, an enclosed firing rod and spring, and three safety devices	
	Sights	Resemble those of the M16-series rifle—	
		Front	Rifle-type, three posts (a central post for engaging stationary or moving targets head-on or straight away, and side posts for engaging targets moving left or right)
		Rear	Sight blade, range adjustment knob, range scale, 2-mm peephole for normal daylight visibility conditions, 7-mm peephole for limited visibility conditions, and 400-meter range indicator, graduated in 50-meter increments
		Night	Can be fitted with the AN/PAQ-4C, AN/PEQ-2, AN/PVS-4, or AN/PAS-13 using fixed mounting rails
	Shoulder Stop	Folding metal and plastic bracket designed to rest on the firer's shoulder to support the launcher while aiming and firing *Note.* When not in use, it is snapped to the underside of the launcher.	
	Sling	Adjustable for carrying the munition	
	Safeties	*Note.* The munition cannot be fired until all three safeties have been disengaged.	
		Transport Safety Fork	Blocks the firing pin from striking the cartridge percussion cap
		Cocking Lever	Has two positions: SAFE and cocked • When in the SAFE position, there is no engagement between the firing rod and the trigger. • When the munition is cocked, the firing rod is engaged with the trigger.
		Safety Release Catch	Connected to the firing rod, prevents the firing rod from striking the firing pin when disengaged *Note.* To disengage the safety release catch, press it down and hold.
	Trigger Assembly	To initiate launch, the firer must press the red trigger button located on the left side of the launcher in tandem with the red safety release catch.	
	Front and Rear Bumpers	Designed to absorb the shock of daily handling and transport, have environmental fire-through muzzle covers	
	Front Grip	Folding front grip used to support the munition while aiming and firing	
COLOR/ MARKINGS	Olive drab background with black-yellow-black color-coded band (HE antiarmor round) or gold color-coded band (FHT)		
	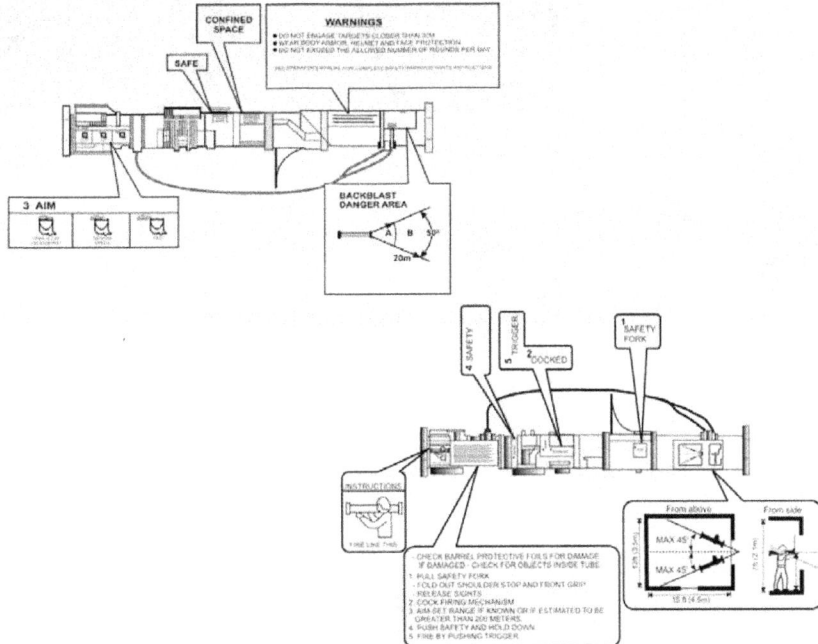		

Table 1-2. Launcher specifications (continued).

		M72A2/A3	
LENGTH	Closed	25 inches	
	Extended	35 inches	
WEIGHT (TOTAL)	M72A2/A3	5 pounds	
COMPONENTS	Launch Tube	Two telescoping aluminum and fiberglass tubes, one inside the other	
		Outer Tube	Fiberglass-reinforced plastic
		Inner Tube	Aluminum
	Firing Mechanism	Mechanical, consists of a firing pin rod with spring, three safety elements, and trigger button	
	Sights	Front	Reticle graduated in 25-meter range increments (50 to 350 meters), curved stadia lines, lead indicators on each side of the stadia lines
		Rear	Steel bracket with a rubber boot and plastic peep sight that adjusts automatically to temperature change
		Night	Can be fitted with the AN/PAQ-4C, AN/PEQ-2, AN/PVS-4, or AN/PAS-13 using the nightsight mounting bracket (NSN 5340-01-391-3004)
	Shoulder Stop	Folding metal rear cover designed to rest against the firer's shoulder to support the launcher while aiming and firing	
	Sling	Adjustable for carrying the munition and providing firing support	
	Safeties	Transport Safety Pin	Keeps the launcher from extending and secures the rear cover to the launcher **Note.** The launcher will not arm if the tube is not fully extended and locked.
		Trigger Safety Handle	Located on the top rear of the outer tube, just forward of the trigger spring boot; pulled to release, and pushed in to prevent ignition
	Trigger Assembly	Trigger spring boot located on top of the launcher, just between the rear sight and the trigger safety handle	
	Front End Cover and Rear Cover	Must be removed before firing, protect the rocket against dirt and moisture	
		Front	Attached to the launcher, acts as shoulder stop while aiming and firing
		Rear	Attached to the sling, which is removed for firing
		Note. Do not discard the sling; it must be replaced if the munition is not fired.	
COLOR/ MARKINGS	Olive drab/black background with white printed label (ROCKET HE 66 MM ANTITANK M72A2/A3 W/COUPLER), practice trainers will be labeled as such in place of "HE ROCKET" 		

Table 1-2. Launcher specifications (continued).

M72A4/5/6/7 (IMPROVED M72)			
LENGTH	Closed	31 inches	
	Extended	39 inches	
WEIGHT (TOTAL)	M72A4/5/6/7	8 pounds	
COMPONENTS	Launch Tube	Two telescoping aluminum and fiberglass tubes, one inside the other	
		Outer Tube	Fiberglass-reinforced plastic
		Inner Tube	Aluminum
	Firing Mechanism	Mechanical, consists of a firing pin rod with spring, three safety elements, and trigger button	
	Sights	Front	Sight blade, center post, and two lead posts; contains an automatic temperature-compensating element to adjust the front sightpost height to match the rocket performance
		Rear	Sight blade, range adjustment knob, range scale, 2-mm peephole for normal daylight visibility conditions, and 7-mm peephole for limited visibility conditions; 350-meter range indicator, graduated in 50-meter increments
		Night	Can be fitted with the AN/PAQ-4C, AN/PEQ-2, AN/PVS-4, or AN/PAS-13 using the nightsight mounting bracket (NSN 5340-01-391-3004) *Note.* The M72A7 has a fixed mounting rail at the forward end for laser light device use.
	Shoulder Stop	Folding metal and rubber rear cover designed to rest against the firer's shoulder to support the launcher while aiming and firing	
	Sling	Adjustable for carrying the munition and providing firing support	
	Safeties	Transport Safety Pin	Keeps the launcher from extending and secures the rear cover to the launcher *Note.* The launcher will not arm if the tube is not fully extended and locked.
		Trigger Safety Handle	Located on the top rear of the outer tube, just forward of the trigger spring boot; pulled to release and pushed in to prevent ignition
	Trigger Assembly	Trigger spring boot located on top of the launcher, just between the rear sight and the trigger safety handle	
	Front End Cover and Rear Cover	Must be removed before firing, protects the rocket against dirt and moisture	
		Front	Attached to the sling, which is removed for firing
		Rear	Attached to the launcher, acts as shoulder stop while aiming and firing
		Note. Do not discard the sling; it must be replaced if the munition is not fired.	
COLOR/ MARKINGS	Olive drab/black background with white-printed label (ROCKET, HE 66MM M72A4/5/6/7) and black color-coded band, practice trainers will be labeled as such in place of "HE ROCKET" 		

ROCKET

1-6. Each launcher contains a rocket, which is propelled from the launcher upon ignition. Table 1-3 depicts the rockets associated with shoulder-launched munitions.

Note. The weights and measurements are approximate. See the corresponding TM for more information.

Table 1-3. Rocket specifications.

M141 BDM	
HE Dual-Mode Assault Rocket	
The dual-mode rocket consists of three major components: a HE warhead, a dual-mode fuze, and a rocket motor. The warhead's function (quick or delay mode) is automatically determined by the fuze when the rocket impacts a target. This automatic feature ensures that the most effective kill mechanism is employed.	
CALIBER	83-mm
MUZZLE VELOCITY	217 meters per second (712 feet per second)
LENGTH	22 inches
WEIGHT	10 pounds
M136 AT4	
High-Explosive Antitank (HEAT) Cartridge	
The M136 AT4 is issued a round of ammunition with an integral, rocket-type cartridge. The cartridge consists of a fin assembly with a tracer element; a point-initiating, base-detonating, piezoelectric fuze; a warhead body with liner; and a precision shaped explosive charge.	
CALIBER	84-mm
MUZZLE VELOCITY	290 meters per second
LENGTH	18 inches
WEIGHT	4 pounds
M136A1 AT4CS	
HEAT Cartridge	
The M136A1 AT4CS is issued a round of ammunition with an integral, rocket-type cartridge. The cartridge consists of a fin assembly with a tracer element; a point-initiating, base-detonating, piezoelectric fuze; a warhead body with liner; and a precision shaped explosive charge.	
CALIBER	84-mm
MUZZLE VELOCITY	225 meters per second
LENGTH	18 inches
WEIGHT	4 pounds

Table 1-3. Rocket specifications (continued).

M72A2/A3	
HE Rocket	

The M72A2/A3 is issued with a round of ammunition. It contains a nonadjustable propelling charge and a rocket. Every M72A2/A3 has an integral HEAT warhead in the rocket's head or body section. The fuze and booster are in the rocket's closure section. The propellant, its igniter, and the fin assembly are in the rocket's motor. No inert versions are available.

CALIBER	66-mm
MUZZLE VELOCITY	145 meters per second
LENGTH	20 inches
WEIGHT	2 pounds

M72A4/5/6/7 (IMPROVED M72)	
HE Rocket	

The round of ammunition is issued with the improved M72.

M72A4:   The warhead is composed of aluminum and is filled with 70/30 octol explosive. It also has a lightened igniter body mass for reduced rear debris danger zone. The warhead also contains a M412A1 fuze modified for a higher velocity rocket.
M72A5:   The warhead is the same as the M72A4, but has a M72A3 ogive to provide armor penetration.
M72A6:   The warhead is the same as the M72A4, but has a copper explosively formed penetrator (EFP) liner for penetration with larger diameter holes and greater spallation effects.
M72A7:   The warhead is the same as the M72A6, but is filled with PBXN-9 explosive. The warhead also has a graze function.

CALIBER	66-mm
MUZZLE VELOCITY	200 meters per second
LENGTH	21 inches
WEIGHT	3 pounds

Effects

1-7. While the operations of shoulder-launched munitions are similar, they produce different effects (Table 1-4).

Table 1-4. Effects of shoulder-launched munitions.

M141 BDM—HE DUAL-MODE ASSAULT ROCKET	
The 83-mm HE assault rocket warhead consists of a dual-mode fuze, an aluminized composition A-3 explosive charge, and 2.38 pounds of explosive.	
Impact/Ignition	Warhead detonation is instantaneous when impacting a hard target, such as a brick or concrete wall, or an armored vehicle. Impact with a softer target, such as a sandbagged bunker, results in a fuze time delay that permits the rocket to penetrate the target before warhead detonation.
Penetration	Penetration of a soft target is enhanced by the high kinetic energy retained by the rocket as it impacts the target. The rocket motor case is located directly behind the warhead, providing additional energy to drive the warhead into the target. The rocket configuration also provides directional stability as the rocket enters soft targets, which greatly enhances lethality, especially when engaging targets at oblique angles. This directional stability after impact keeps the rocket from deflecting away from the target wall.

M136 AT4 and M136A1 AT4CS—HEAT CARTRIDGE

The M136 AT4's warhead has excellent penetration ability and lethal after-armor effects. The extremely destructive, 440-gram shaped-charge explosive penetrates about 14 inches (M136 AT4) or 16 inches (M136A1 AT4CS) of armor.

IMPACT   IGNITION   PENETRATION   SPALLATION

Impact	The nose cone crushes; the impact sensor activates the fuze.
Ignition	The piezoelectric fuze element activates the electric detonator. The booster detonates, initiating the main charge.
Penetration	The main charge fires and forces the warhead body liner into a directional gas jet that penetrates armor plate.
After-Armor Effects (Spallation)	The projectile fragments and incendiary effects produce blinding light and highly destructive results.

M72A2/A3 and M72A7—66-mm HE ROCKET

The 66-mm HEAT rocket warhead consists of a tapered, thin-gauge steel body. Once it explodes, the force and heat of the explosive focus into a small, but powerful, gas jet. This directional jet penetrates the target and, if the target is a vehicle, sprays molten metal inside. If the jet hits an engine or ammunition, it may start a fire or cause an explosion.

IMPACT   IGNITION   PENETRATION   SPALLATION

Impact	The nose cone crushes; the impact sensor activates the fuze.
Ignition	The ogive crush switch activates the electric detonator. The booster detonates, initiating the main charge.
Penetration	The main charge fires and forces the warhead body liner into a directional gas jet that penetrates armor plate.
After-Armor Effects (Spallation)	The projectile fragments and incendiary effects produce blinding light and highly destructive results.

OBSERVATION DEVICES

1-8. Shoulder-launched munitions can have one or two types of sights:
- Fixed launcher sights.
- Attachable sights.

Fixed Launcher Sights

1-9. Shoulder-launched munition sights come fixed to the round of munition, and are stowed within the launcher and released for use. Except for the M72-series, all shoulder-launched munition sights are similar in design. Each sight requires its own methods of ranging and sighting a target.

M136-Series Shoulder-Launched Munitions and M141 Bunker Defeat Munitions

1-10. M136-series shoulder-launched munitions and M141 BDM launchers have front and rear blade sights (Figure 1-2).

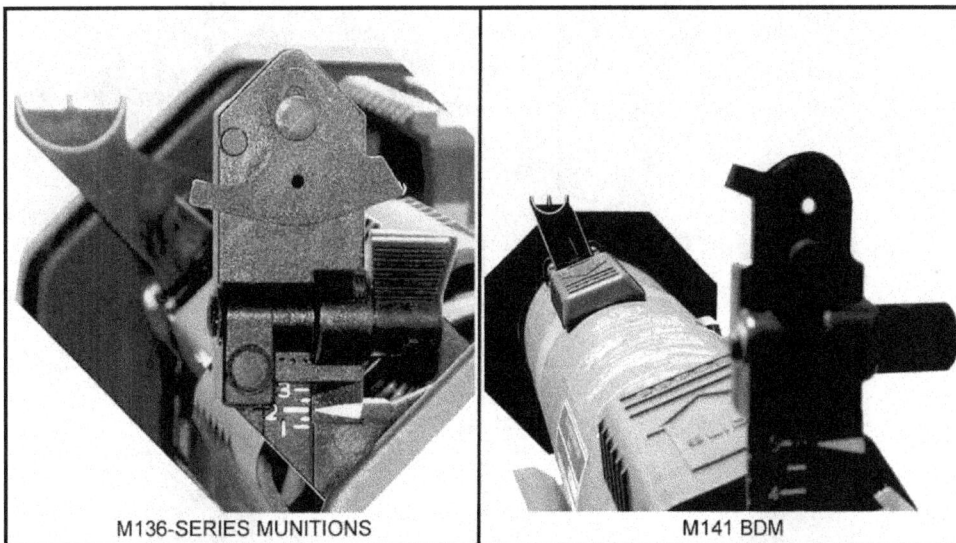

M136-SERIES MUNITIONS | M141 BDM

Figure 1-2. M136-series shoulder-launched munitions and M141 bunker defeat munition front and rear blade sights.

Front Sight

1-11. The front sight (Figure 1-3) has a sight blade with a center post and left and right lead posts. A semicircular white line helps the firer obtain the proper sight picture.

Rear Sight

1-12. The rear sight (Figure 1-4) has a sight blade, range adjustment knob, range scale, 2-mm peephole for normal daylight visibility conditions, and 7-mm peephole for limited visibility conditions. The range scale is indexed with ranges from 100 to 500 meters in 50-meter increments, with the exception of the M136A1 AT4CS, which is indexed from 100 to 400 meters.

> *Note.* When firing M136-series munitions and the M141 BDM, the range should be set to the nearest 50 meters and not be automatically left on the battlesight setting (200 meters for M136-series munitions, 150 meters for the M141 BDM) to avoid missing the target and causing possible duds.

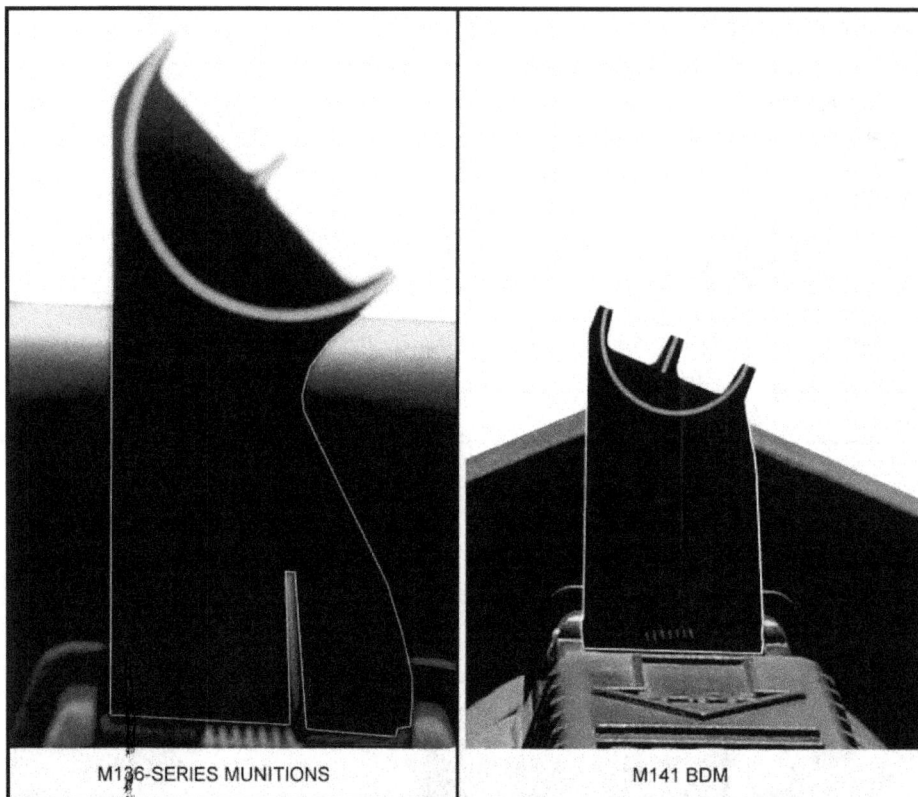

Figure 1-3. M136-series shoulder-launched munitions and
M141 bunker defeat munition front sight.

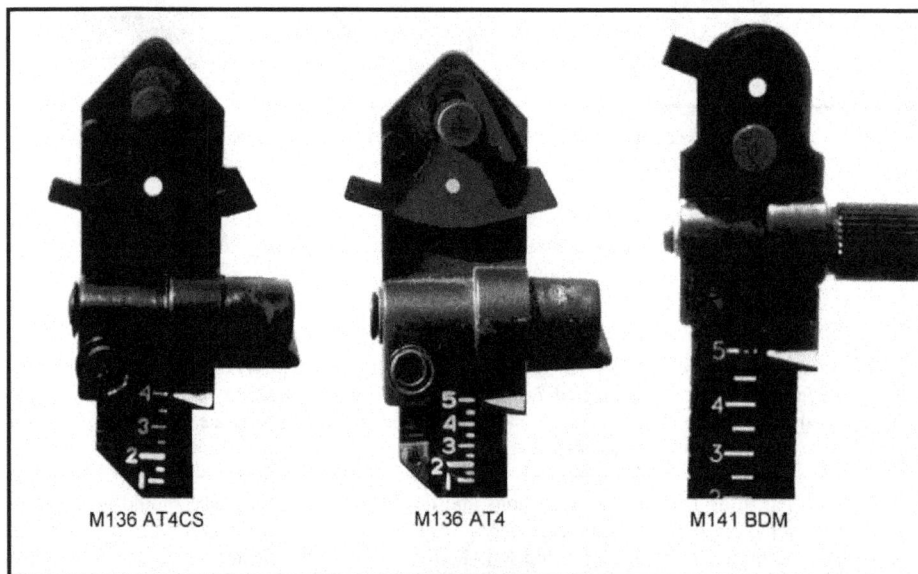

Figure 1-4. M136-series shoulder-launched munitions and
M141 bunker defeat munition rear sight.

M72A2/A3 Shoulder-Launched Munitions

1-13. The M72A2/A3 has front and rear sight blade sights (Figure 1-5).

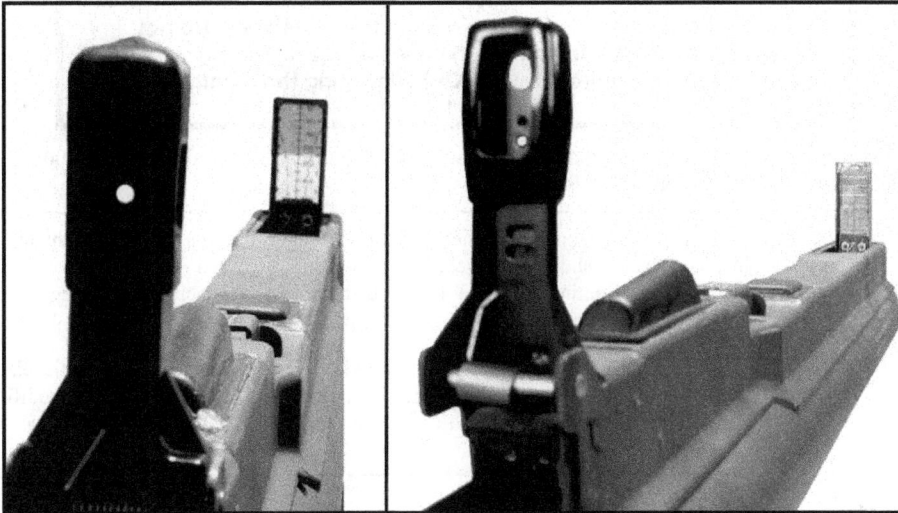

Figure 1-5. M72A2/A3 front and rear blade sights.

Front Sight

1-14. The M72A2 launcher front sight (Figure 1-6) has a raised vertical range line marked with ranges from 50 to 350 meters in 25-meter increments. Two curved stadia lines are etched on the front sights. Lead indicators are located on either side of the stadia lines to help engage moving targets.

Note. Do not use the stadia lines on this sight to estimate range, because they are inaccurate.

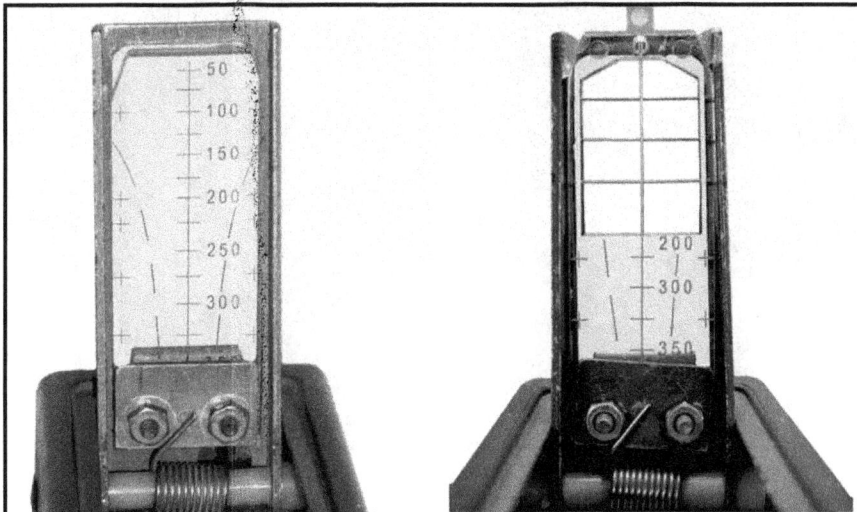

Figure 1-6. M72A2/A3 front sight.

WARNING

M72A2/A3 front sights contain a radioactive substance. Do not handle excessively. Detach and dispose of in accordance with the standing operating procedure (SOP) after firing the munition.

1-15. On the M72A3, Soldiers should use the front sight illuminated range marks at the 100- and 150-meter points to help engage targets in low light.

Notes.	1.	Not all M72A3 shoulder-launched munitions have illuminated front sight range marks. Because shoulder-launched munitions are discarded after firing in combat, sights with illuminated range marks that contain radioactive substances were eventually discontinued due to environmental concerns.
	2.	To resolve complaints that the firer couldn't see through the clear plastic blade during limited visibility conditions if it was dirty, the front sight was also modified to include a simple wire lattice in the top half of the front sight picture (Figure 1-6).

Rear Sight

1-16. M72A2/A3 launchers have the same rear sight. The rear sight (Figure 1-7) consists of a steel bracket with a rubber boot and plastic peep sight. This sight automatically adjusts to changes in temperature; its settings are unaffected by temperature.

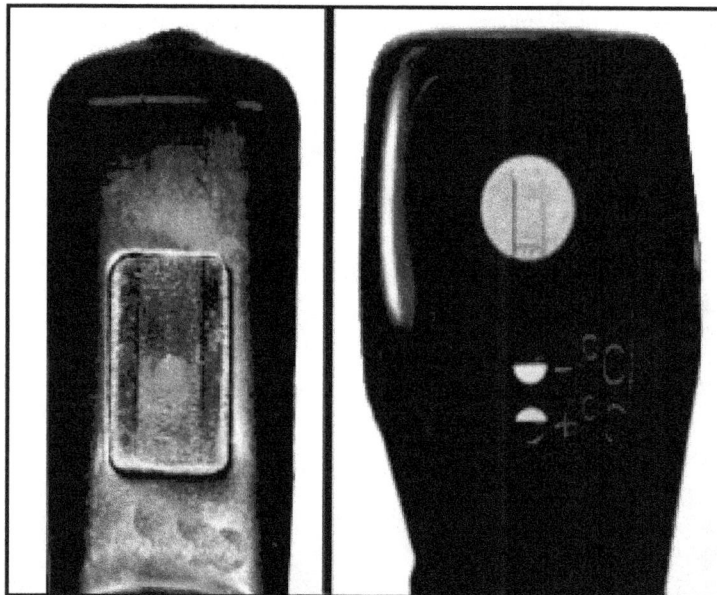

Figure 1-7. M72A2/A3 rear sight.

Improved M72 Shoulder-Launched Munitions

1-17. Improved M72 shoulder-launched munitions launchers have front and rear blade sights (Figure 1-8).

**Figure 1-8. Improved M72 shoulder-launched munitions
front and rear blade sights.**

Front Sight

1-18. The front sight (Figure 1-9) has three lead posts to help line up fast-moving, slow-moving, or stationary targets. The front sight is spring-loaded to automatically adjust for temperature-induced performance differences.

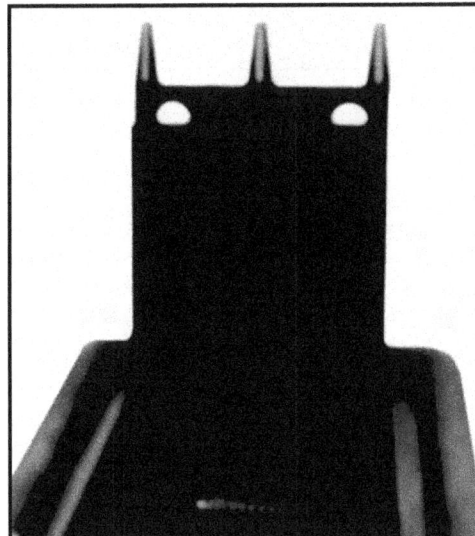

Figure 1-9. Improved M72 front sight.

Rear Sight

1-19. The rear sight (Figure 1-10) is more like a standard rifle sight; it has a range setting knob, a range indicator in 50-meter increments, and two apertures (peep holes), a daylight aperture and a low light aperture.

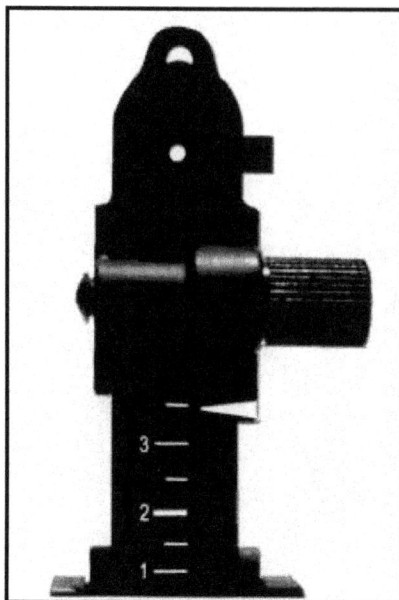

Figure 1-10. Improved M72 rear sight.

Attachable Sights

1-20. Shoulder-launched munitions do not come with dedicated night vision sights (NVSs). Soldiers must use unit weapon night vision devices (NVDs) when conducting operations during limited visibility conditions. External sights, such as the AN/PVS-4 and AN/PAS-13-series NVSs, and the AN/PAQ-4 and AN/PEQ-series laser aiming lights/illuminators can be mounted directly to the M136A1 AT4CS, the M141 BDM, and the improved M72. The M72A2/A3 and the M136 AT4 require a NVD mounting bracket.

Note. The NVD mounting bracket kit is issued with the AN/PVS-4 NVS.

AN/PVS-4 Night Vision Sight

1-21. The AN/PVS-4 NVS (Figure 1-11) is issued with various accessories, including a bracket that allows for use on the M72A2/A3 and M136 AT4. However, this works only if field maintenance has already installed a M72A1 (for M72A2/A3) or M67 (for the M136 AT4 and the M72 improved launcher) reticle.

Note. Though the reticle was developed for the M72A1, it can also be used with M72A2/A3 launchers.

Figure 1-11. AN/PVS-4 night vision sight.

AN/PAS-13B/C/D (V1) Light Weapon Thermal Sight and AN/PAS-13B/C/D (V3) Heavy Weapon Thermal Sight

1-22. The AN/PAS-13B/C/D (V1) light weapon thermal sight (LWTS) and the AN/PAS-13B/C/D (V3) heavy weapon thermal sight (HWTS) (Figure 1-12) are silent, lightweight, compact, and durable battery-powered infrared (IR) imaging sensors that operate with low battery consumption.

Notes.	1.	Both the LWTS and the HWTS are referred to henceforth as a singular thermal weapon sight (TWS). For more information, refer to TMs 11-5855-312-10, 11-5855-316-10, and 11-5855-317-10.
	2.	The TWS can be used on all shoulder-launched munitions, but the LWTS is recommended to reduce launcher weight.

1-23. The TWS is capable of target acquisition under conditions of limited visibility, such as darkness, smoke, fog, dust, and haze. It operates effectively both at night and during the day. The TWS is composed of two functional groups: the telescope and the basic sensor.

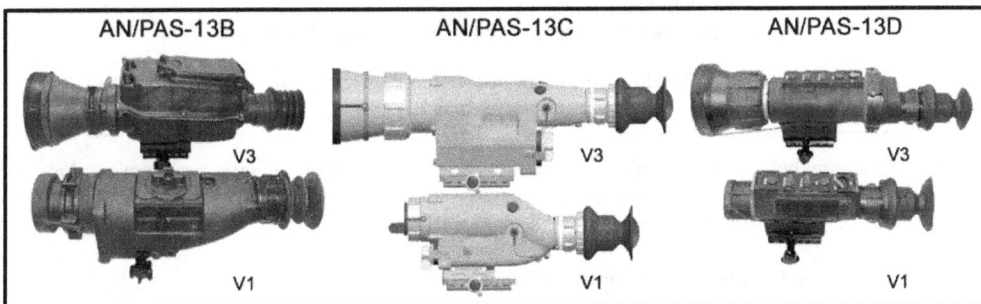

Figure 1-12. AN/PAS-13B/C/D (V1) light weapon thermal sight
and AN/PAS-13B/C/D (V3) heavy weapon thermal sight.

AN/PAQ-4B/C Infrared Aiming Light

1-24. The AN/PAQ-4B/C IR aiming light (Figure 1-13) projects an IR laser beam that is invisible to the naked eye, but can be seen with NVDs. This aiming light works with the AN/PVS-7-series goggles and the AN/PVS-14.

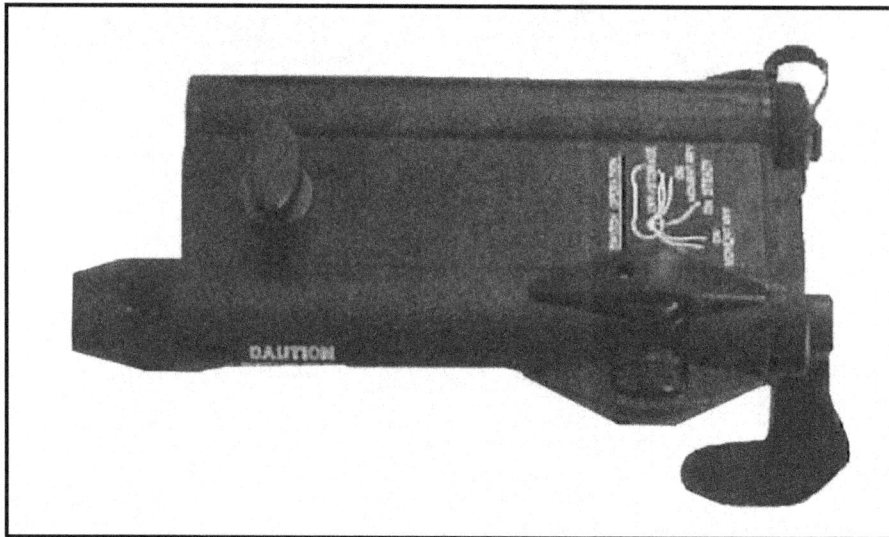

Figure 1-13. AN/PAQ-4B/C infrared aiming light.

AN/PEQ-2A/B Target Pointer/Illuminator/Aiming Light

1-25. AN/PEQ-2A and AN/PEQ-2B aiming lights (Figure 1-14) are Class IIIb laser devices that emit a collimated beam of IR light for precise aiming and a separate IR beam for illumination of the target or target area. Both beams can be independently zeroed to the munition and to each other. The beams can be operated individually or in combination in both high and low power settings.

Notes.	1.	The IR illuminator is equipped with an adjustable bezel to vary the size of the illumination beam based on the size and distance of the target.
	2.	A safety block is provided for training purposes to limit the operator from selecting high power modes of operation.

1-26. The aiming lights are used with NVDs and can be used as handheld illuminators/pointers or mounted on the munition with the included brackets and accessory mounts. In the mounted mode, the aiming lights can be used to direct fire and to illuminate and designate targets.

AN/PEQ-15 Advanced Target Pointer/Illuminator/Aiming Light

1-27. The AN/PEQ-15 advanced target pointer/illuminator/aiming light (Figure 1-15) is a multifunctional laser device that emits visible or IR light for precise weapon aiming and target/area illumination. It has two different types of lasers:

- The visible aiming laser provides for active target acquisition in low-light and close quarters combat situations without the need for NVDs.
- The IR aiming and illumination lasers provide for active, covert target acquisition in low light or complete darkness when used in conjunction with NVDs.

Note. The AN/PEQ-15 can be used as either a handheld illuminator/pointer or can be mounted to weapons equipped with a MIL-STD-1913 rail.

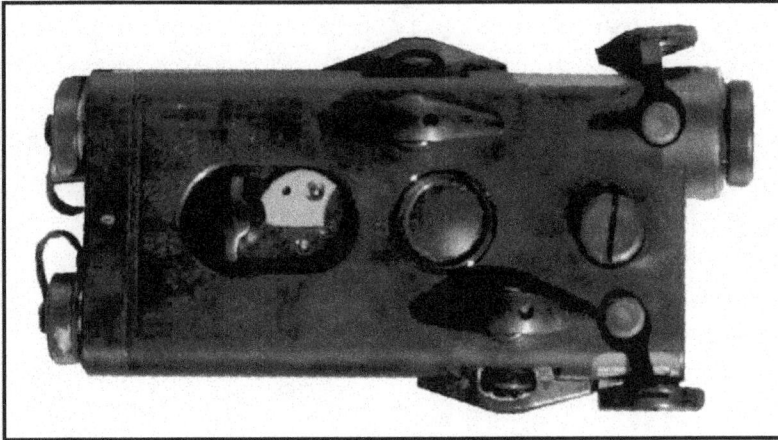

Figure 1-14. AN/PEQ-2A/B target pointer/illuminator/aiming light.

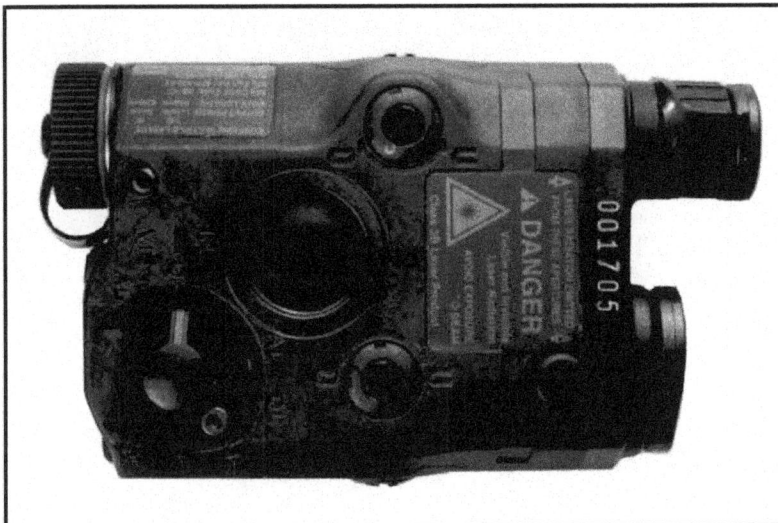

Figure 1-15. AN/PEQ-15 advanced target pointer/illuminator/aiming light.

This page intentionally left blank.

Chapter 2
TRAINING

An effective training strategy integrates resources into a year-round program to train the individual and collective skills needed to perform the unit's wartime mission. This ensures that units are trained to fight and win on the battlefield. The training strategy supports both the generating force and operational Army.

SECTION I. TRAINING STRATEGY

The Army shoulder-launched munition training strategy is a concept of integrating resources into a program to train and sustain individual and collective marksmanship skills. The shoulder-launched munition training strategy begins in initial entry training (IET) and continues to the unit, where sustainment and collective training continue to build on the basic skills and introduce additional skills.

OBJECTIVES

2-1.   The procedures and techniques for implementing the Army shoulder-launched munition training strategy are based on the concept that designated Soldiers should understand common firing principles and be confident in their ability to apply their firing skills in combat.

OVERVIEW

2-2.   There are two primary components of a marksmanship training strategy: initial training and sustainment training. Both may include individual and collective tasks and skills.

Note. If a long period of time elapses between initial and sustainment sessions or training doctrine is altered, retraining may be required.

INITIAL TRAINING

2-3.   In IET, Soldiers learn about the operation and function of shoulder-launched munitions using a hands-on training approach with the field handling trainer (FHT) or field-expedient trainer (FET). The tasks include the following:

- Receive an orientation safety briefing.
- Perform pre-fire serviceability checks on a shoulder-launched munition.
- Prepare a shoulder-launched munition for firing.
- Demonstrate correct firing positions.
- Determine correct sight picture.
- Understand and apply the fundamentals of marksmanship:
  - Steady hold.
  - Aiming procedures, including eye placement, sight alignment, and sight picture.
  - Breath control.
  - Trigger manipulation.
- Perform misfire procedures.
- Restore the shoulder-launched munition to a carrying configuration.

2-4.   IET training culminates in the Soldier's proficiency assessment, in which the Soldier demonstrates the integrated act of firing using the M287 subcaliber training launcher. This evaluation enables leaders to determine the effectiveness of the training. Figure 2-1 shows the IET training strategy.

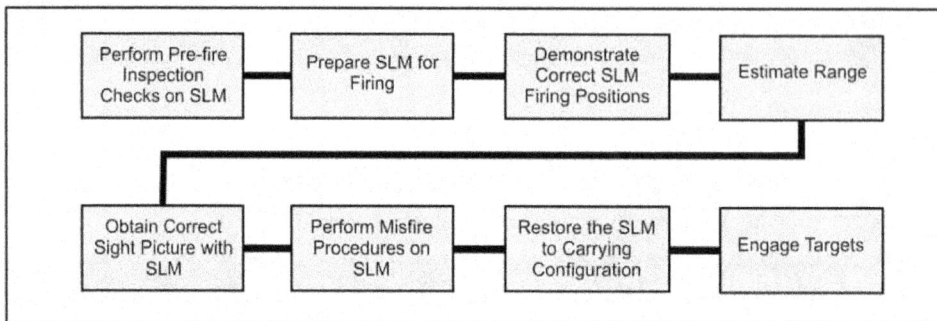

Figure 2-1. Training strategy for initial entry training.

SUSTAINMENT TRAINING

2-5.   Training continues in regular Army, Army National Guard (ARNG), and United States Army Reserve (USAR) units using the same basic skills taught in IET.

Note. Units must have a plan not only for when they are at their home station, but for when they are deployed as well.

2-6.   To sustain the basic marksmanship skills taught in IET, training is conducted, followed by practice and qualification fire. Key elements include—

- The training of trainers.
- The use of training aids, devices, simulators, and simulations (TADSS).
- Sustainment training.
- Remedial training.

Note. See Appendix B for more information about TADSS.

2-7.   Additional skills trained in the unit include—

- Types of targets.
  - Tracked and wheeled vehicles.
  - Manmade structures.
- Target engageability.
  - Estimate range.
  - Stationary and moving target engagement techniques.
  - Munition selection.
  - Attack points.
- Firing while wearing mission-oriented protective posture (MOPP) gear.
- Firing using aiming devices and NVDs, including—
  - Mounting.
  - Operating.
  - Sight alignment.
  - Target engagement.
- Limited visibility training.

2-8.   These skills are trained and integrated into collective training exercises, such as squad and platoon field training exercises (FTXs) and live-fire situational training exercises (STXs).

2-9.   General marksmanship knowledge and weapon proficiency are perishable skills. A year-round marksmanship sustainment program is needed for the unit to maintain the individual and collective firing proficiency requirements to accomplish its mission. Figure 2-2 shows a year-round training strategy guide.

> *Note.* Currently, no single type of shoulder-launched munition can eliminate both armor and reinforced, hardened targets. Because of this, Soldiers must be familiar with the purpose and employment of all types of shoulder-launched munitions.

DESIGNATED MARKSMEN TRAINING

2-10.   In addition to the conduct of a year-round training strategy for all Soldiers, commanders should select three Soldiers per squad to serve as the primary and alternate shoulder-launched munition designated marksmen.

2-11.   These Soldiers should receive sustainment training more frequently and be expected to maintain a higher level of proficiency; they should conduct sustainment training quarterly and fire the appropriate tables semiannually. These Soldiers should also receive more frequent opportunities to fire live munitions.

> *Notes.*   1.   Subcaliber training launchers enable Soldiers to practice applying the fundamentals of marksmanship, but do not fully prepare Soldiers for the blast effects of live munitions. These blast effects can affect the Soldier's accuracy, and designated marksmen should become accustomed to these effects so that they can place accurate fire. For this reason, annual firing with live munitions is recommended.
>
> 2.   Soldier accuracy deteriorates after experiencing the blast effects of the initial round. Firing assessments prove that blast anticipation after firing the initial round causes the firer to concentrate more on blast effects than the target. This can be overcome if Soldiers are given the opportunity to fire more shoulder-launched munitions and at a greater frequency. Soldiers can use simulators that closely replicate the blast effects of firing live munitions to reduce firer anticipation.

2-12.   Commanders should use the unit's experienced designated marksmen as trainers and range safety personnel when conducting shoulder-launched munition training. When working as an assistant instructor, these designated marksmen can load subcaliber training launchers to reduce time on the firing line and enforce safety procedures.

Selection Criteria

> *Note.* Soldiers designated to carry or fire a shoulder-launched munition at the squad or platoon can be of any rank as long as they meet the other criteria.

2-13.   When selecting designated marksmen, commanders should consider—
- Longevity.
- Qualification scores.

Longevity

2-14.   Commanders should consider using newly assigned Soldiers to reduce retraining due to personnel turnover.

Qualification Scores

2-15.   The fundamentals of rifle marksmanship are similar to those of shoulder-launched munition marksmanship.

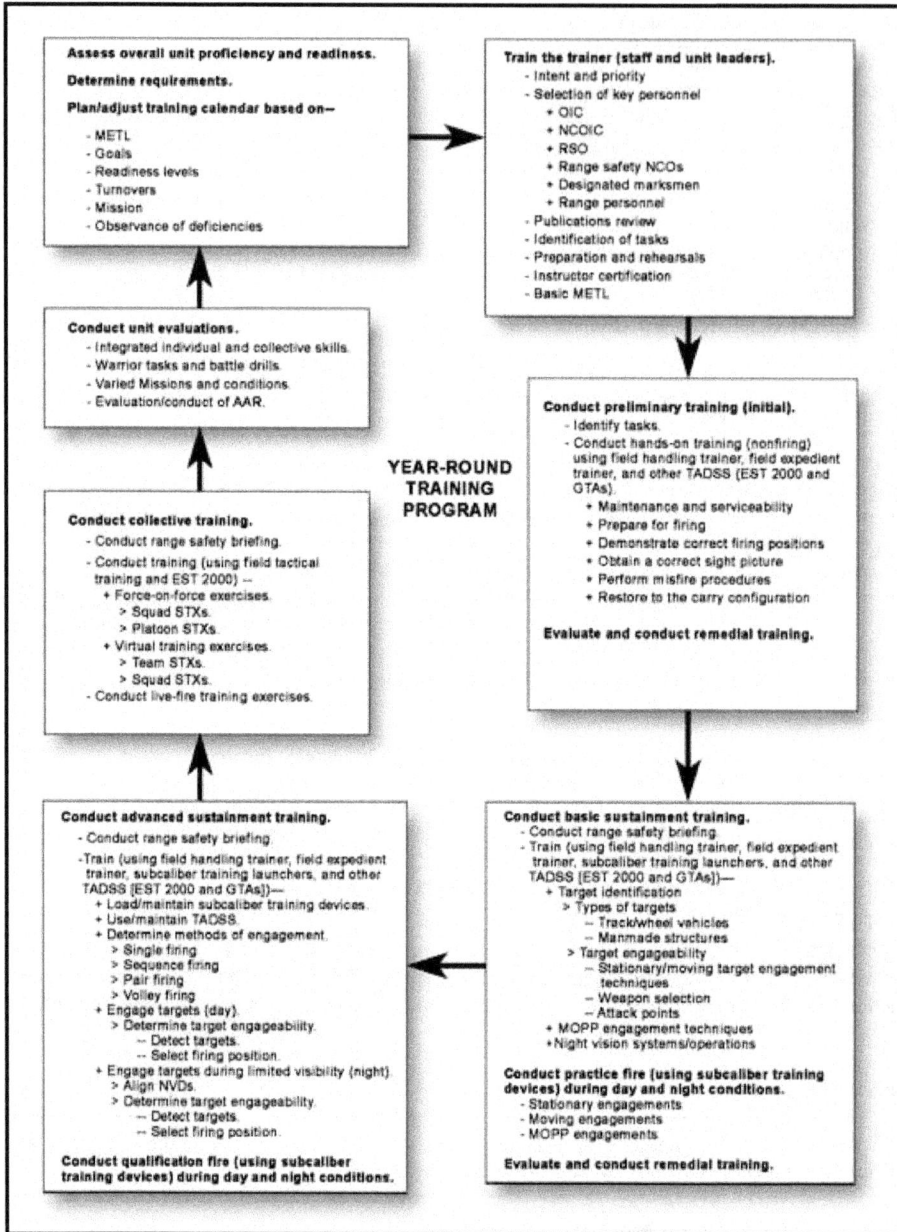

Figure 2-2. Unit marksmanship sustainment strategy.

QUALIFICATION REQUIREMENTS

2-16. Qualification requirements for Soldiers assigned to an Infantry or reconnaissance platoon/section differ from those for Soldiers assigned to maneuver and sustainment units.

Note. See AR 350-1 for specific requirements pertaining to marksmanship training and DA PAM 350-38 for live-fire frequency requirements.

TRAINING PHASES

2-17. Soldiers progress through three phases of training:
- Preliminary.
- Basic.
- Advanced.

PRELIMINARY

2-18. Preliminary shoulder-launched munition marksmanship training covers those tasks learned during a Soldier's initial training, as well as target engagement procedures. Preliminary marksmanship training includes the following tasks:
- Perform serviceability checks on a shoulder-launched munition.
- Prepare a shoulder-launched munition for firing.
- Demonstrate correct firing positions.
- Estimate range to a target.
- Apply the fundamentals of marksmanship (to obtain a correct sight picture), such as—
  - Steady hold.
  - Aiming procedures, including eye placement, sight alignment, and sight picture.
  - Breath control.
  - Trigger manipulation.
- Perform misfire procedures.
- Return the shoulder-launched munition to the carrying configuration.

Notes.   1.   Preliminary tasks should be taught or reinforced before conducting any form of live-fire training.

2.   With proper training and oversight by the instructor/trainer, a Soldier with poor marksmanship skills can improve those skills with the help of the Engagement Skills Trainer (EST) 2000. See Appendix B for more information about the EST 2000.

BASIC

2-19. Basic shoulder-launched munition marksmanship training begins at the conclusion of Phase I. Basic shoulder-launched munition marksmanship training includes the following tasks:
- Identify targets.
- Determine target engageability.
- Engage targets using MOPP engagement techniques (practice day fire).
- Install and operate NVSs.
- Install and operate aiming lights.
- Perform practice fire (using subcaliber training launchers) in daytime and limited visibility conditions.
- Use other TADSS.

ADVANCED

2-20.   Advanced shoulder-launched munition marksmanship training begins at the conclusion of Phase II. Advanced shoulder-launched munition marksmanship training includes the following tasks:
- Load and maintain the subcaliber training launchers.
- Use and maintain other TADSS.
- Use the proper methods of engagement (single, sequence, pair, and volley fire).
- Perform qualification fire (using subcaliber training launchers) in daytime conditions.

- Use and maintain NVSs, and perform sight alignment procedures.
- Use and maintain aiming lights, and perform sight alignment procedures.
- Perform qualification fire (using subcaliber training launchers) in limited visibility conditions.

COLLECTIVE

2-21. Collective training begins at the conclusion of Phase III. Collective training includes—

- Perform collective squad and platoon training exercises.
- Conduct live-fire training exercises.

SECTION II. UNIT MARKSMANSHIP TRAINING PROGRAM

2-22. An effective unit marksmanship program reflects the priority, emphasis, and interest of commanders and trainers. This section outlines a marksmanship training program strategy as guidance in establishing and conducting an effective unit training program. The strategy consists of the individual and leader refresher training for maintaining the basic skills learned during IET. It progresses to training advanced and collective skills.

MISSION-ESSENTIAL TASKS

2-23. Unit commanders should focus their shoulder-launched munition training programs to support their mission-essential task lists (METLs).

Note. Refer to FM 7-0 for more information about developing METLs and long-range, short-range, and near-term training plans.

TRAINING ASSESSMENT

2-24. To conduct an effective shoulder-launched munition marksmanship program, the unit commander must determine the current marksmanship proficiency of all assigned personnel. Constant evaluation provides commanders understanding of where training emphasis is needed. All results are reviewed to determine any areas that need strengthening, along with any individuals that require special attention. Based on this evaluation, marksmanship training programs are developed and executed. Commanders continually assess the program and modify it as required. To develop a training plan and assess the marksmanship program, commanders should use the following tools:

- Direct observation of training.
- Spot checks.
- Review of past training.

2-25. Based on the commander's evaluation, goals, and missions, quarterly, semiannual, or annual training events are identified.

DIRECT OBSERVATION OF TRAINING

2-26. Observing and accurately recording performance reveals each Soldier's practice and qualification fire results and ability to hit targets. This also enables the commander to identify Soldiers who need special assistance to reach required standards and those who exceed these standards.

SPOT CHECKS

2-27. Spot checks of individual marksmanship performance, such as interviews and evaluations of Soldiers, provide commanders with valuable information about Soldier proficiency and knowledge of the marksmanship tasks.

REVIEW OF PAST TRAINING

2-28. Commanders review past training to gain valuable information for developing a training plan. The assessment should include the frequency and results of training.

COMMANDER'S EVALUATION GUIDE

2-29. The commander's evaluation guide contains three sections:
- Commander's priorities and intent.
- Soldier assessment.
- Trainer assessment.

2-30. The following is an example of a commander's evaluation guide. Commanders can use this guide not only to assess their unit's marksmanship proficiency, but also to assess the unit leaders and their ability to effectively implement a marksmanship program. They can also use it to develop noncommissioned officers (NCOs) into subject matter experts.

Commander's Priorities and Intent

2-31. When considering their priorities and intent, commanders answer the following questions:
- Have you clearly stated the priority of shoulder-launched munition proficiency in your unit? What is it? Do the staff and subordinates support this priority? Is it based on your METL and an understanding of FM 7-0?
- Have you clearly stated that practice and qualification fire are opportunities for the commander to assess several skills relating to shoulder-launched munition readiness?
- How will practice and qualification fire be conducted? Will the prescribed procedures be followed? Who will collect the data?
- Have you clearly stated the purpose and intent of preliminary instruction?
    - What skills will preliminary instruction address?
    - Will preliminary instruction be performance-oriented? Are tasks integrated?
- Have you determined the designated marksmen in your unit? How many? Who are they? What additional training will they receive? How and where? What resources will this training require?

Soldier Assessment

2-32. During Soldier assessment, commanders answer the following questions:
- Do Soldiers maintain the munition in accordance with the TM? Do they have manuals?
- Do Soldiers conduct serviceability checks of the munition before training? Were deficiencies noted?
- Do Soldiers demonstrate an understanding of the munition's operation, functioning, and capabilities?
- Can Soldiers correctly apply misfire procedures? Have they demonstrated this during dry-fire exercises?
- Can Soldiers precisely and consistently apply the fundamentals of shoulder-launched munition marksmanship? To what standard have they demonstrated their mastery?
    - During preliminary training?
    - During sustainment training?
    - During a dry-fire exercise?
    - During a live-fire exercise (LFX)?
    - During a collective training exercise?
- Can Soldiers correctly mount external sighting systems to shoulder-launched munitions (NVS/aiming light) and accurately align the devices to standard?
    - Do they understand how to properly mount and operate the NVS and aiming light?
    - Do they understand sight alignment procedures for the NVS and aiming light?
    - Do they understand sight alignment procedures?

- Do Soldiers demonstrate their knowledge of the effects of movement, wind, and gravity while firing? What feedback was provided? How?
- Can Soldiers scan a designated area and detect all targets out to the maximum range of shoulder-launched munitions? If not, why?
- Can Soldiers perform range estimation? If not, why?
- Can Soldiers quickly engage targets from all firing positions out to the maximum effective range of shoulder-launched munitions? If not, which targets were not engaged? Which were missed? Why?
- What is the hit distribution during collective LFXs?
- Do Soldiers demonstrate proficiency during target detection and acquisition techniques? When using NVDs?
- Do Soldiers demonstrate individual marksmanship proficiency while wearing MOPP gear? During collective exercises?
- Do Soldiers demonstrate proficiency during moving target engagements? If not, is moving target training conducted?
- Based on onsite observations and analysis of training and firing performance, what skills or tasks show a readiness deficiency?
  - What skills need training emphasis? Individual emphasis? Leader emphasis?
  - What are the performance goals?

Trainer Assessment

2-33. During trainer assessment, commanders answer the following questions:

- Who has trained or will train the trainers?
  - What is the subject matter expertise of the cadre?
  - Are they actually training the critical skills?
  - What aids and devices are used? Is EST 2000 properly used?
- What administrative constraints or training distracters can you overcome for the junior officer and NCO?
- At what level are the resources necessary to train marksmanship controlled (time, training aids, munitions, ranges)?
- Do the sergeants perform the duties they are charged with?

INSTRUCTORS/TRAINERS

2-34. Knowledgeable instructors or cadre are the key to marksmanship performance. All commanders must be aware of maintaining expertise in marksmanship instruction/training in accordance with general subject technical manuals (GS TMs), field manuals (FMs), TMs, Army regulations (ARs), and command SOPs.

SELECTION

2-35. Trainers within a unit are normally team, squad, and section leaders and platoon sergeants. These Soldiers must—

- Demonstrate proficiency in all aspects of shoulder-launched munition marksmanship.
- Be proficient in applying the fundamentals.
- Know the importance of marksmanship training.
- Have qualified with all live munitions.
- Demonstrate competence and a professional attitude.

2-36. Before becoming trainers, they must be assessed carefully and their shortcomings must be corrected. The commander chooses a method of assessing the trainers that ensures that their abilities are accurately evaluated. With the assistance of unit senior trainers (command sergeants major and company first sergeants), platoon leaders, and platoon sergeants, the commander performs the assessment.

DUTIES

2-37. Instructors/trainers help firers master the fundamentals of shoulder-launched munition marksmanship. They ensure that firers consistently apply what they have learned. They must also perform the following tasks:

- Set up and run a range.
- Conduct an orientation safety briefing.
- Inspect the munitions for serviceability.
- Prepare the munitions for firing.
- Demonstrate the correct firing positions.
- Estimate range.
- Obtain the correct sight picture.
- Perform the correct combat and training misfire procedures.
- Return the munition to the carrying configuration.
- Mount NVDs and conduct sight alignment procedures.
- Set up and fire subcaliber training launchers.
- Coach marksmanship techniques.

2-38. Successful trainers know how to operate the training devices for the shoulder-launched munitions assigned to their units. Trainers must know the appropriate combat techniques for employing shoulder-launched munitions.

TRAINING THE TRAINER

2-39. The goal of a progressive train-the-trainer program is to achieve a high state of combat readiness. Through the active and aggressive leadership of the chain of command, a perpetual base of expertise is established and maintained.

Note. The commander should identify unit personnel who have had assignments as marksmanship instructors. These individuals should be used to train other unit cadre by conducting preliminary, basic, and advanced marksmanship instruction for their Soldiers.

2-40. A suggested train-the-trainer program is outlined below:

- Conduct the marksmanship diagnostic test.
- Review operation and function of the munition, including misfire and safety procedures.
- Review effects of wind, gravity, and movement when firing.
- Review coaching techniques and device usage.
- Diagnose firing problems.
- Conduct preliminary, basic, and advanced marksmanship instruction.
- Conduct range operations.
- Conduct practice and qualification fire.
- Conduct shoulder-launched munitions live-fire training exercises.
- Plan and conduct squad STXs and platoon FTXs.

TRAINER CERTIFICATION PROGRAM

2-41. The certification program sustains the trainers' expertise and develops methods of training. The program standardizes procedures for certifying marksmanship trainers. Trainers' technical expertise must be continuously refreshed, updated, and closely managed.

2-42. The training base can expect the same personnel changes as any other organization. Soldiers assigned as marksmanship trainers will have varying degrees of experience and knowledge of training procedures and methods. Therefore, the trainer certification program must be an ongoing process that is tailored to address these variables.

2-43. All marksmanship trainers must complete the four phases of training using the progression steps, and the records of training should be updated on a quarterly basis.

Notes.   1.   Trainers who fail to attend or do not pass any phase of the diagnostic examination will be assigned to subsequent training.

           2.   At a minimum, formal records should document program progression for each trainer.

Phase I

2-44. During this phase, the trainer must accomplish the following tasks and be certified by the chain of command:

- Be briefed on the concept of the certification program.
- Be briefed on the unit marksmanship training strategy.
- Review the unit marksmanship training outlines.
- Review issued reference material.
- Visit training sites and firing ranges.

Phase II

2-45. Phase II should be completed no more than two weeks following the conclusion of Phase I. During Phase II, the trainer demonstrates his mastery of all skills taught during preliminary, basic, and advanced marksmanship training, and his performance is reviewed by the chain of command. The results of this review are recorded and maintained on the trainer's progression sheet, which is, in turn, designed in accordance with the unit SOP.

Phase III

2-46. During this phase, the trainer sets up and conducts firing on the various ranges. He explains the targets and scoring procedures. The trainer explains the purpose of practice and qualification fire, range layout, and the conduct of training. This validates that the trainer has gained the knowledge necessary to conduct training. The results of this review are recorded and maintained on the trainer's progression sheet.

Phase IV

2-47. The final phase of the train-the-trainer program tests the trainer. During this phase, the trainer sets up a range and conducts training for at least one person. If ammunition and subcaliber training launchers are available, the trainer conducts a firing exercise. If ammunition and subcaliber training launchers are not available, the evaluation is based on the quality of training given.

SECTION III. TRAINING PREPARATION

2-48. Training preparation involves three steps:

     (1) Conduct a training risk assessment.
     (2) Conduct an environmental risk assessment.
     (3) Make range coordinations.

CONDUCT A TRAINING RISK ASSESSMENT

2-49. The officer in charge (OIC) or noncommissioned officer in charge (NCOIC) conducts a training risk assessment. It is vital to identify unnecessary risks by comparing potential benefit to potential loss. The composite risk management (CRM) process allows units to identify and control hazards, conserve combat power and resources, and accomplish the mission. This process is cyclic and continuous; it must be integrated into all phases of operations and training.

> Application of the risk management process will not detract from this training goal, but will enhance execution of highly effective, realistic training.
>
> FM 7-0, TRAINING FOR FULL-SPECTRUM OPERATIONS

2-50. There are five steps to the CRM process:

(1) Identify hazards.

(2) Assess hazards to determine risk.

(3) Develop control measures and make risk decisions.

(4) Implement control measures.

(5) Supervise and evaluate.

Note. Risk decisions must be made at the appropriate level.

IDENTIFY HAZARDS

2-51. When identifying hazards, leaders should consider—

- The lethality of the shoulder-launched munitions used.
- The area in which training is to be conducted.
- How the addition of new elements impacts known hazards.

Surface Danger Zones

2-52. Surface danger zones (SDZs) are exclusion areas identified to protect personnel from the munitions fired during training. Each SDZ contains two areas:

- Backblast danger area.
- Downrange danger area.

Backblast Danger Area

2-53. When all shoulder-launched munitions are fired, propellant gases exit from the back of the launcher with tremendous force. The resulting backblast (heat, overpressure, and launch debris) can damage equipment or seriously injure personnel who are too close to the rear of the launcher.

> **DANGER**
>
> **DURING TRAINING, THE ENTIRE BACKBLAST AREA MUST BE MARKED OFF AND KEPT CLEAR OF PERSONNEL, EQUIPMENT, AND OBSTRUCTIONS.**

M141 Bunker Defeat Munition

2-54. The backblast danger area for the M141 BDM is similar in composition and characteristics to that of the M136 AT4, but contains an expanded ear protection caution area. The munition produces sound pressure levels that may exceed 140 decibels; dangerous noise levels exist within 445 meters of a fired munition (Figure 2-3).

Note. See TM 9-1340-228-10 for more information.

DANGER

WHEN THE M141 BDM IS FIRED, EAR PROTECTION MUST BE WORN IN THE EAR PROTECTION CAUTION AREA. PERMANENT EAR DAMAGE MAY RESULT IF YOU FIRE THE M141 BDM WITHOUT EAR PROTECTION OR YOU ARE WITHIN 445 METERS LEFT, RIGHT, OR TO THE REAR OF THE FIRED LAUNCHER WITHOUT EAR PROTECTION.

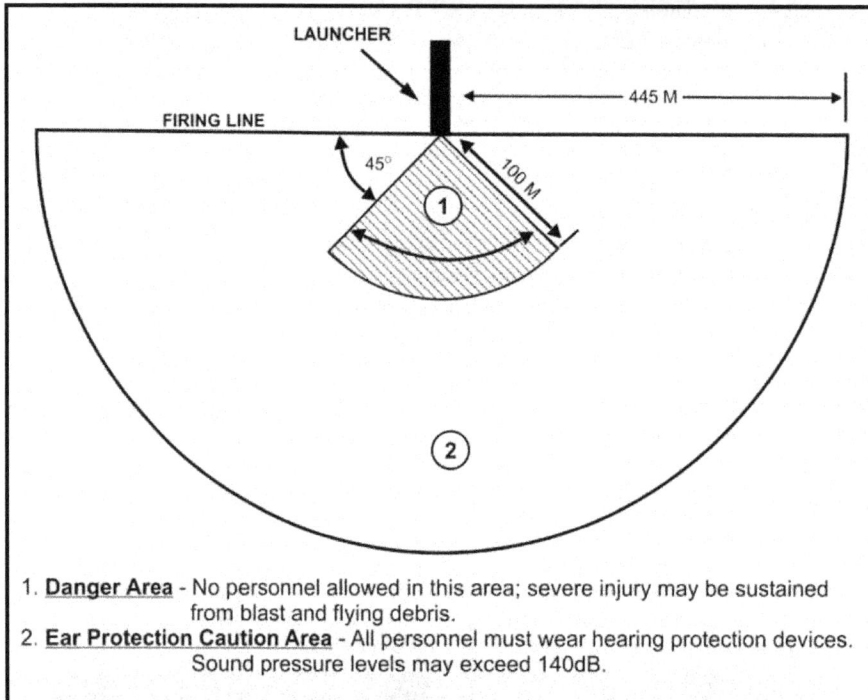

LAUNCHER

445 M

FIRING LINE

45°

100 M

1

2

1. **Danger Area** - No personnel allowed in this area; severe injury may be sustained from blast and flying debris.
2. **Ear Protection Caution Area** - All personnel must wear hearing protection devices. Sound pressure levels may exceed 140dB.

Figure 2-3. M141 bunker defeat munition backblast and ear protection caution area.

M136 AT4

2-55. The total backblast area extends 100 meters to the rear of the launcher in a 90-degree fan (Figure 2-4).

WARNINGS

This munition must not be fired from an enclosure or from a fighting or prone position during training. See TM 9-1315-886-12 for more information.

When operating temperatures fall below freezing (0 degrees Celsius or 32 degrees Fahrenheit), the dimensions of all backblast areas and safety zones double.

M136Al AT4 Confined Space

2-56. The M136Al AT4CS backblast area contains two danger zones (Figure 2-5):

- Danger Zone A.
- Danger Zone B.

2-57. Danger Zone A presents dangers from backblast, heat, and flying debris. Obstacles such as barriers, big trees, or other vertical objects must not be in this zone. Danger Zone B presents dangers from backblast, flying debris, and noise.

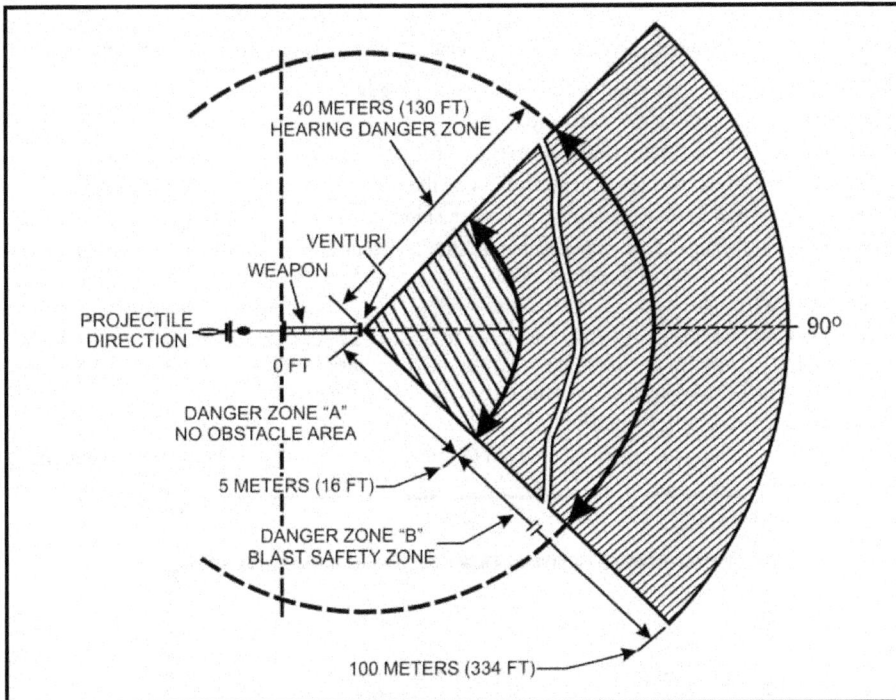

Figure 2-4. M136 AT4 backblast area.

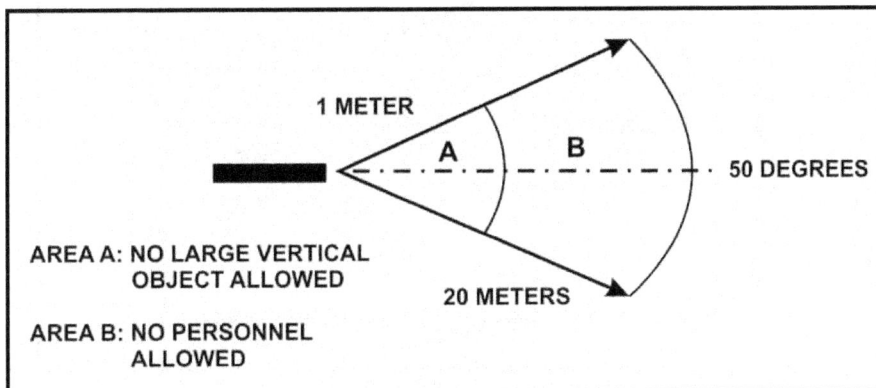

Figure 2-5. M136A1 AT4 confined space backblast area.

Firing From a Confined Area

2-58. The confined space must be a room with the following dimensions or larger (Figure 2-6):

- The inside area must be a minimum of 12 feet wide and 15 feet long (about 3.5 meters wide and 4.5 meters long).
- The ceiling must be a minimum of 7 feet (2.1 meters) high.
- The window opening must be a minimum of 36 inches wide and 36 inches long (1 meter wide and 1 meter long).
- The door opening must be a minimum of 36 inches wide and 72 inches long (1 meter wide and 2 meters long).
- The structure should be of significant construction to withstand the munition's backblast.

2-59. The following requirements must be followed when firing indoors:

- Fire in the standing position only.
- Cover and/or protect all equipment (i.e., small arms, radio set, etc.) in the room.
- Remove any loose objects which might be thrown when firing from directly behind the launcher.
- Keep stuffed furniture (e.g., mattresses, cushions, pillows, etc.) in the room to absorb pressure.
- Hang a blanket 1.5 to 2 meters behind the launcher and 15 to 30 cm from the rear wall. This considerably reduces sound pressure.
- Open all windows and doors in room.
- Do not allow the angle of the launcher to exceed 20 degrees of depression from the horizontal plane. Do not fire the munition at any angle of elevation (Figure 2-7). Do not allow the angle of the launcher to exceed 45 degrees left or right from the vertical plane.
- Wear combat arms earplugs (CAEs).
- Fire the munition no more than 10 cm (4 inches) from a door or window frame (Figure 2-8).

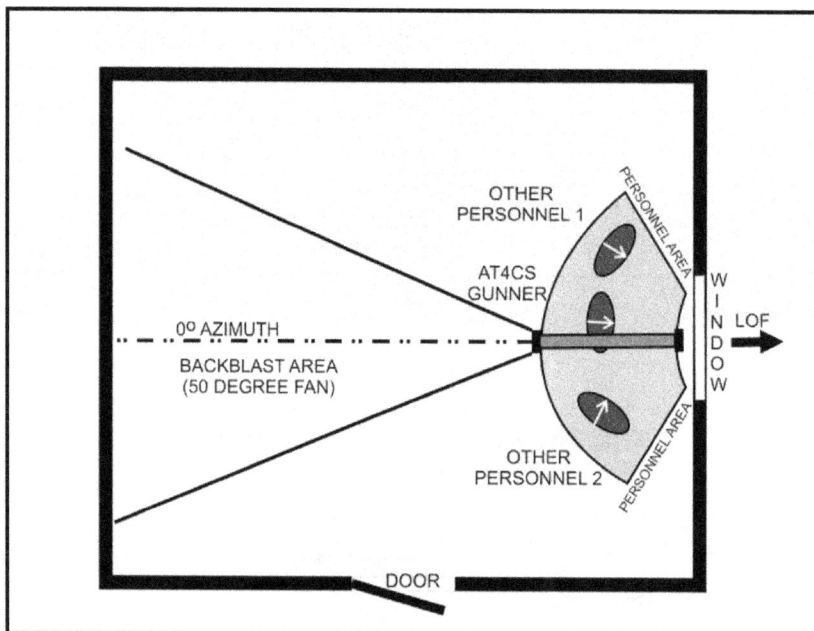

Figure 2-6. Minimal dimensions of a confined space.

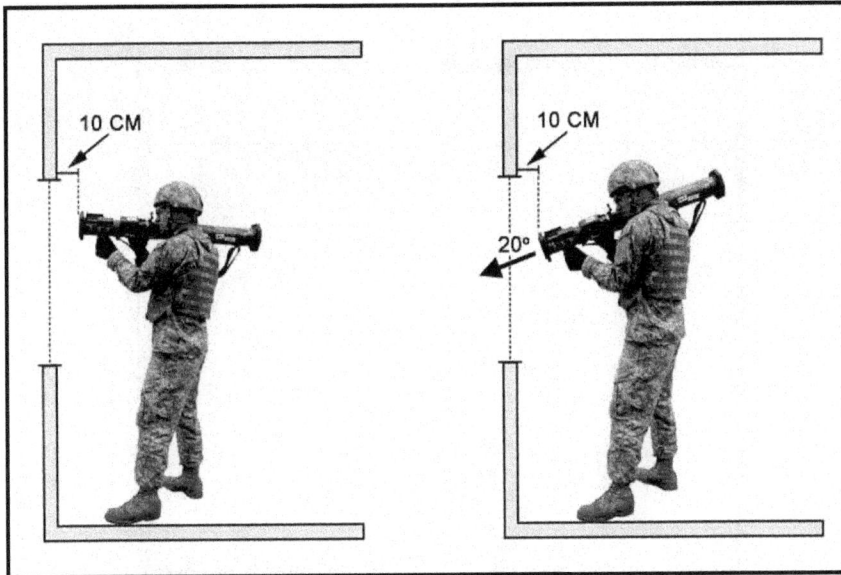

Figure 2-7. Angle of launcher.

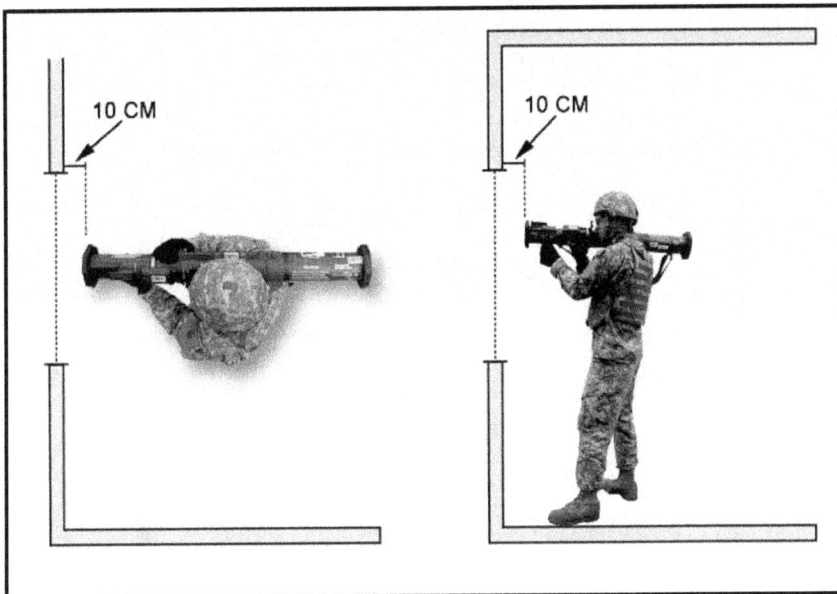

Figure 2-8. Minimal distance from door or window frame.

2-60. The following restrictions apply when firing indoors with additional personnel (more than the firer) occupying the room:

- No more than three personnel (including the firer) are allowed inside the room.
- Soldiers providing cover fire for the firer must be positioned outside of the munition's backblast area (Figures 2-9 and 2-10).

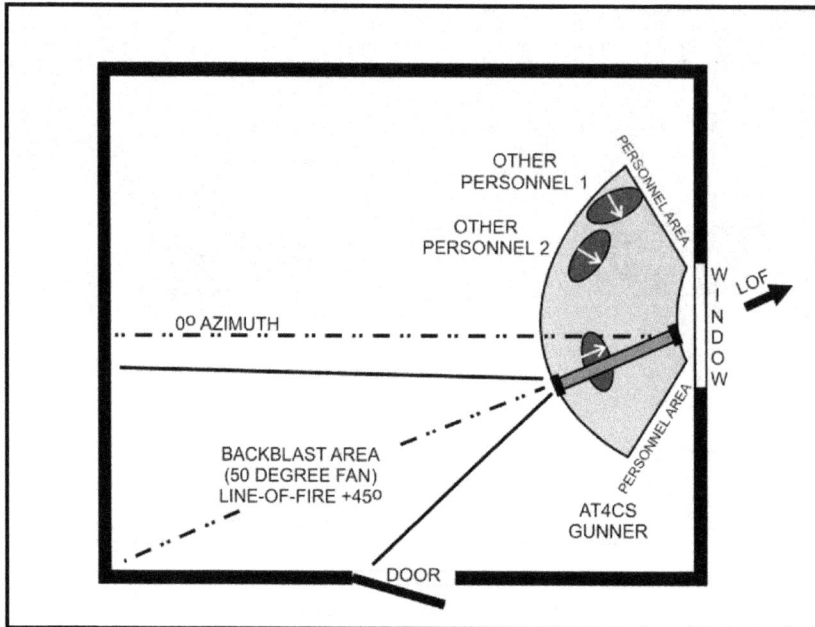

Figure 2-9. Soldier positions for firing a M136A1 AT4 confined space on an oblique left azimuth.

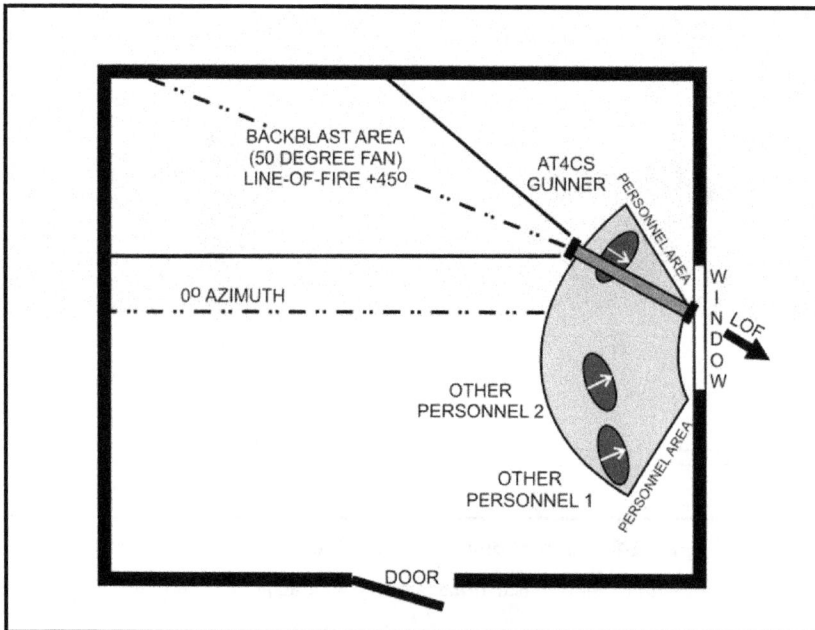

Figure 2-10. Soldier positions for firing a M136A1 AT4 confined space on an oblique right azimuth.

M72A2 and M72A3

2-61. The total backblast area extends 40 meters (44 yards) to the rear of the launcher and is divided into two zones (Figure 2-11):

- Danger zone.
- Caution zone.

Note. During training, both zones should be marked off limits.

2-62. All personnel, equipment, and flammable material must be clear of the danger zone. The munition's backblast may throw loose objects to the rear; therefore, personnel must also stay clear of the caution area.

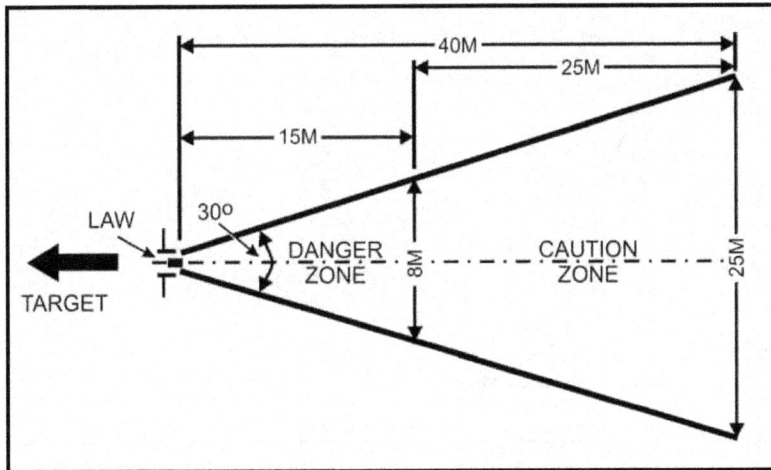

Figure 2-11. M72A2 and M72A3 backblast area.

M72A4/A5/A6/A7

2-63. The total backblast area for these munitions extends 70 meters to the rear of the launcher (Figure 2-12).

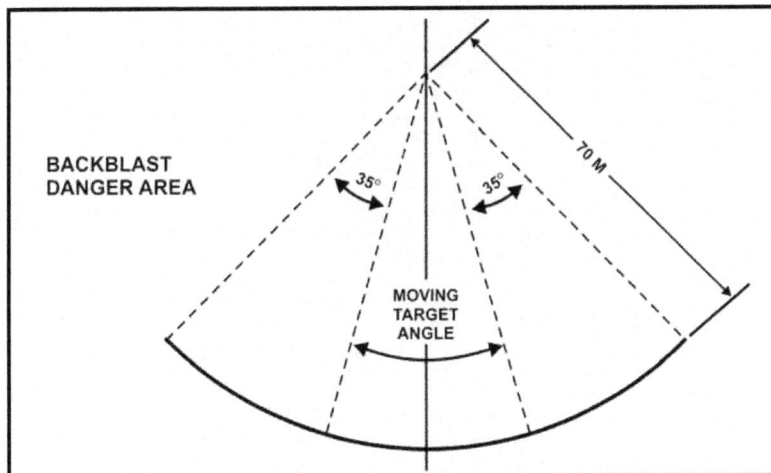

Figure 2-12. M72A4/A5/A6/A7 backblast area.

Note. The total backblast area is greater for the M72A4/A5/A6/A7 than for the M72A2/A3.

Downrange Danger Area

2-64. The downrange danger area is the area in front of the launcher into which the munition will fire.

M141 Bunker Defeat Munition

2-65. The downrange danger area requirements for the M141 BDM are illustrated in Figure 2-13.

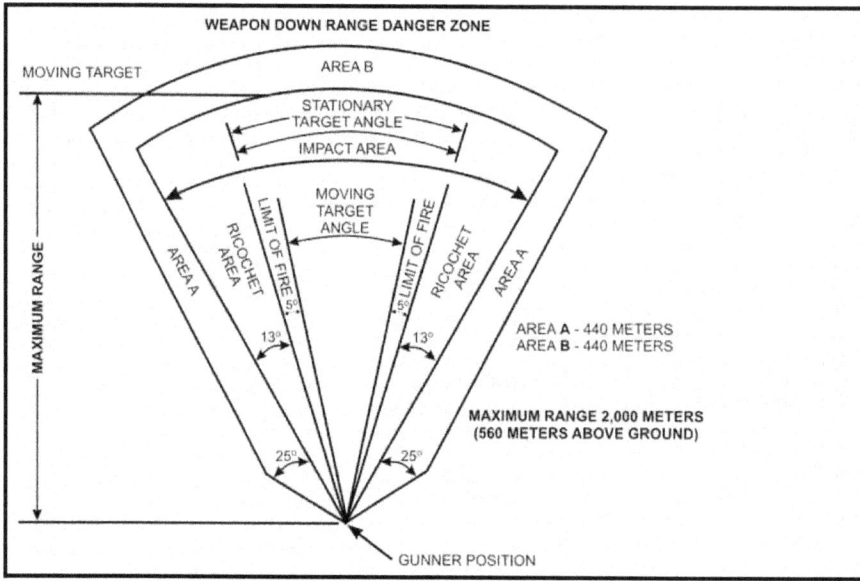

Figure 2-13. Downrange danger area for the M141 bunker defeat munition.

M136 AT4

2-66. The downrange danger area requirements for the M136 AT4 are shown in Figure 2-14.

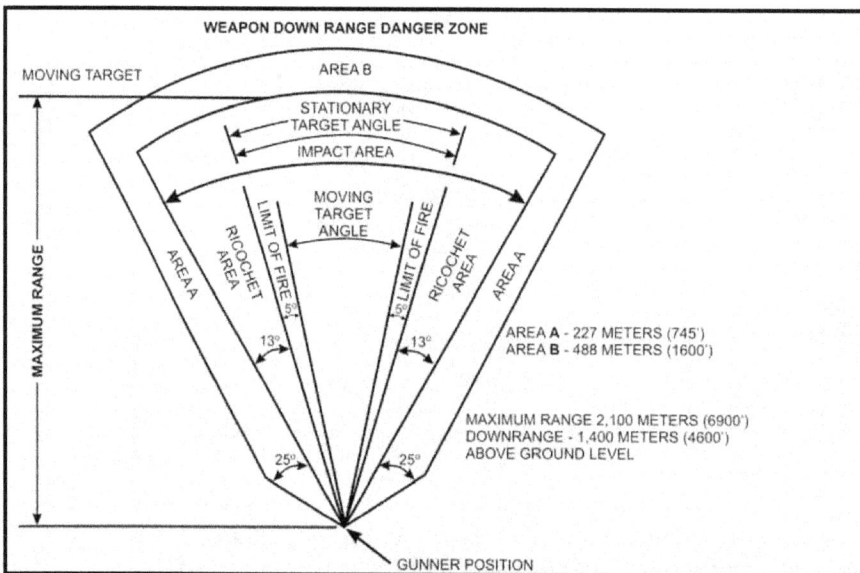

Figure 2-14. Downrange danger area for the M136 AT4.

M136A1 AT4 Confined Space

2-67. Figure 2-15 shows the downrange danger area requirements for the M136A1 AT4CS.

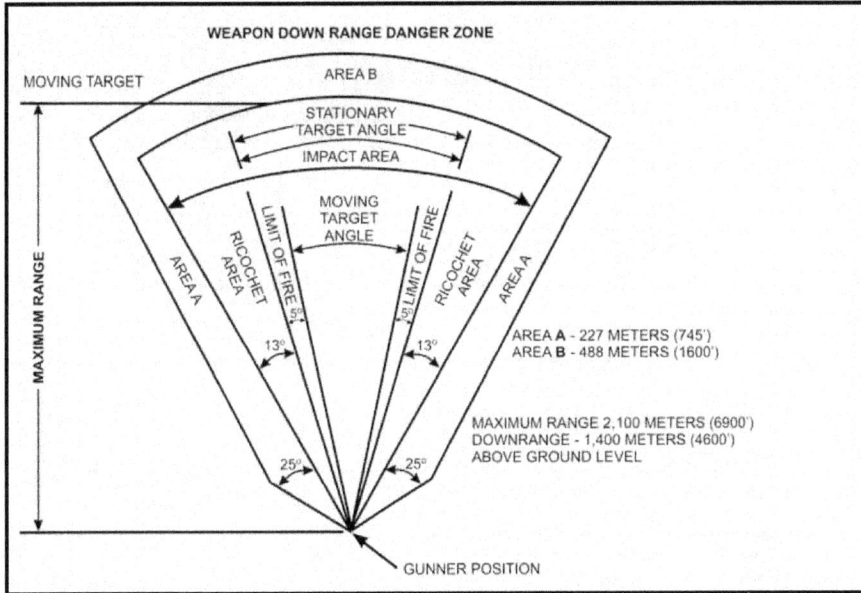

Figure 2-15. Downrange danger area for the M136A1 AT4 confined space.

M72A4/A5/A6/A7

2-68. Figure 2-16 shows the downrange danger area requirements for the M72A4/A5/A6/A7.

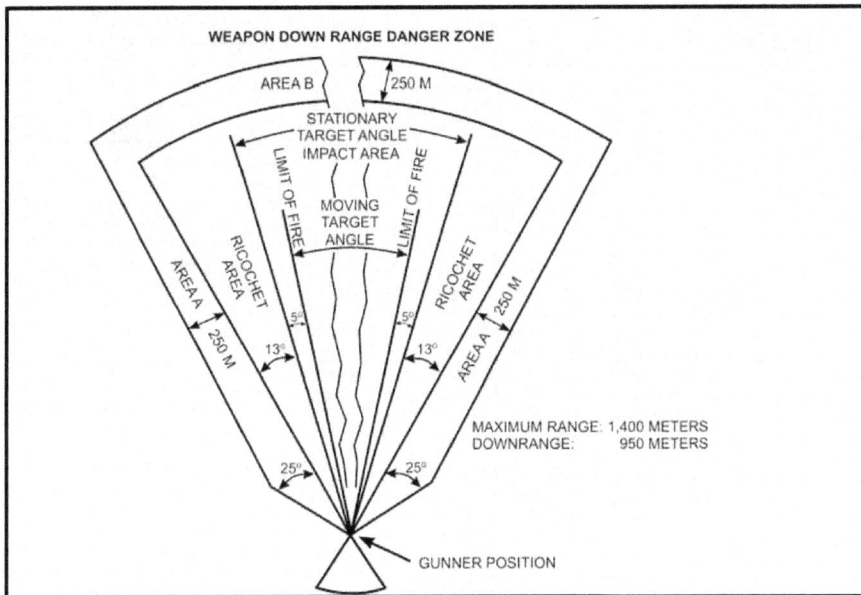

Figure 2-16. Downrange danger area for the M72A4/A5/A6/A7.

Operating Temperatures

2-69. No shoulder-launched munition should be fired when its temperature exceeds its operating temperature limit range (Table 2-1).

CAUTION

Firing shoulder-launched munitions in temperatures outside operating temperature limits could cause a misfire or produce some other hazard for the Soldier.

Note. When operating in cold weather, bringing the M141 BDM into a warm enclosure may cause damage. The change in temperature will make metal components sweat and the moisture can cause rust or corrosion.

WARNING

Do not use sharp instruments to chip off snow or ice. Do not thaw a munition near a direct flame.

Note. In hot weather, munitions should be stored in a shaded area and properly ventilated, whether they are located at training sites or at fighting positions.

Table 2-1. Operating temperatures.

MUNITION	OPERATING TEMPERATURES	
	STORAGE	FIRING
M141 BDM	-50°F (-45°C) to +160°F (+70°C)	-25°F to +120°F (-32°C to +49°C)
M136 AT4	-50°F to +160°F (-46°C to +71°C)	-40°F to +140°F (-40°C to +60°C)
M136A1 AT4CS	-50°F to +160°F (-46°C to +71°C)	-40°F to +140°F (-40°C to +60°C)
M72A4/5/6/7 (Improved M72)	-50°F to +160°F (-45.6°C to +71°C)	-40°F to +140°F (-40°C to +60°C)
M72A2/A3	-40°F to +140°F (-40°C to +60°C)	-40°F to +140°F (-40°C to +60°C)

Target Engagement Ranges

2-70. Fragments formed by the detonation of the tactical warhead can be hazardous to the firer and other personnel who are on, or close to, the firing line. The velocity of these fragments increases with target hardness. Target engagement ranges are designed to place firers where they are safe, but where the targets are still engageable. Table 2-2 shows the target engagement ranges for shoulder-launched munitions.

Table 2-2. Target engagement ranges for shoulder-launched munitions.

MUNITION		M141 BDM	M136 AT4	M136A1 AT4CS	M72A2/A3	Improved M72
Minimum Range	Training	150 meters	160 meters	160 meters	100 meters	100 meters
	Combat	15 meters	10 meters	30 meters	10 meters	25 meters
Arming Range		15 meters	10 meters	9 to 15 meters	9 to15 meters	25 meters
Maximum Range		2,000 meters	2,100 meters	2,100 meters	1,000 meters	1,400 meters
Maximum Engagement Range		500 meters	500 meters	400 meters	250 meters	350 meters
Maximum Effective Range		250 meters	300 meters	300 meters	200 meters for a stationary target, 165 meters for a moving target **Note.** Beyond these ranges, there is less than a 50 percent chance of hitting the target.	220 meters

DANGER

WHEN FIRING SHOULDER-LAUNCHED MUNITIONS WITHIN THE ARMING RANGE, THE FIRER MUST EXPOSE ONLY THOSE PARTS OF THE BODY NEEDED TO ENGAGE THE TARGET. THE FIRER MUST TAKE IMMEDIATE COVER AFTER LAUNCH. ALL FRIENDLY PERSONNEL IN THE AREA OF THE TARGET MUST BE BEHIND PROTECTIVE COVER. TARGET AND ROCKET DEBRIS MAY CAUSE INJURY OR DEATH.

WHEN FIRING A SHOULDER-LAUNCHED MUNITION LESS THAN ITS MINIMUM RANGE, THE FIRER MUST EXPOSE ONLY THOSE PARTS OF THE BODY NEEDED TO ENGAGE THE TARGET. THE FIRER MUST TAKE IMMEDIATE COVER AFTER LAUNCH. ALL FRIENDLY PERSONNEL IN THE AREA OF THE TARGET MUST BE BEHIND PROTECTIVE COVER. TARGET AND ROCKET DEBRIS MAY CAUSE INJURY OR DEATH.

WARNINGS

When firing the M141 BDM at soft targets less than 150 meters (492 feet) away or firing at hard targets less than 200 meters (656 feet) away, fire from behind safe cover (i.e., sandbag wall, concrete wall) to prevent injury from flying debris.

During M136-series munition training, hard targets must be placed at least 160 meters (175 yards) downrange from the firing position. When firing M136-series munitions at targets less than 160 meters away, fire from behind appropriate cover to prevent possible injury from shrapnel. As the target distance decreases, possibility of injury from shrapnel increases.

When firing the M72 at targets less than 100 meters away, fire from behind appropriate cover to prevent possible injury from shrapnel. As target distance decreases, possibility of injury from shrapnel increases.

Dangerous noise levels exist within 445 meters of a fired shoulder-launched munition. Operator and personnel must wear properly fitted, approved earplugs to prevent permanent hearing loss or damage.

A firer of shoulder-launched munitions and all personnel within 100 meters (328 feet) behind a shoulder-launched munition firing line must wear a helmet, protective vest, ear protection, and eye protection or stand behind appropriate safe cover (i.e. sand bag wall or inside a building) to prevent injury from flying debris.

Firing Limitations

2-71. The overpressure and noise created by firing shoulder-launched munitions requires special precautions during training. To reduce these hazards, firers must observe the firing limitations shown in Table 2-3.

DANGER

DO NOT FIRE M136 AT4S, M141 BDMS, OR M72-SERIES MUNITIONS FROM ANY ENCLOSURE OR FROM INDIVIDUAL FIGHTING POSITIONS DURING TRAINING.

WHEN FIRED FROM WITHIN COVERED FIGHTING POSITIONS, THE MUNITION'S BACKBLAST OVERPRESSURE CAN CAUSE STRUCTURAL DAMAGE AND SEROUS INJURY OR DEATH TO THE FIRER AND SOLDIERS PROVIDING SUPPORT.

WHEN FIRING FROM THE MODIFIED STANDING POSITION, THE FIRER MUST KEEP THEIR BACK AGAINST THE WALL OF THE DUG-IN FIGHTING POSITION TO MINIMIZE DEFLECTION. RAISING THE FRONT END OF THE LAUNCHER CAN CAUSE THE BACKBLAST TO BE DEFLECTED ONTO THE FIRER, CAUSING INJURY OR DEATH. IF THE FIGHTING POSITION RESTRICTS THE BACKBLAST AREA, THE FIRER SHOULD MOVE TO AN ABOVE-GROUND POSITION BEFORE FIRING THE SHOULDER-LAUNCHED MUNITION.

Table 2-3. Range firing limitations.

MUNITION	ASSOCIATED MATERIAL	MAXIMUM NUMBER OF ROUNDS THAT MAY BE FIRED IN A 24-HOUR PERIOD		
M141 BDM	TM 9-1340-228-10	Prone		1
		Sitting		0
		Kneeling		3
		Standing		6
M136 AT4	TM 9-1315-886-12	Prone		0
		Sitting		1
		Kneeling		3
		Standing		3
M136A1 AT4CS	TM 9-1315-255-13	Outdoor (with single ear protection)	Prone	70
			Sitting	0
			Kneeling	14
			Standing	28
		Indoor (with combat earplugs)	Prone	0
			Sitting	0
			Kneeling	0
			Standing	1
M72A2/A3	TM 9-1340-214-10	4 (for Soldier firing the munition and personnel within 20 meters of the launcher, given that properly fitted, approved earplugs are worn)		
M72A7	TB 9-1340-230-13			
Note. Shoulder-launched munitions TMs and this manual explain four firing positions: standing, kneeling, sitting, and prone. Although shoulder-launched munitions can be fired from the four positions, the sitting and prone positions increase the chances that blast/overpressure will injure the firer. Soldiers should be trained on assuming the four firing positions, but only live fire from the standing and kneeling positions.				

Duds

2-72. Soldiers should treat duds as hazardous ammunition. They should take corrective actions for removal and disposal of dud rounds of ammunition in accordance with the unit SOP.

ASSESS HAZARDS TO DETERMINE RISK

2-73. Once identified, a hazard is assessed by considering the likelihood of its occurrence and the potential severity of injury without considering any control measures. When assessing hazards, leaders should consider the Soldiers' current state of training.

DEVELOP CONTROL MEASURES AND MAKE RISK DECISIONS

2-74. Leaders must apply two types of control measures to shoulder-launched munition risk assessments:
- Educational control measures.
- Physical control measures.

2-75. The unit commander's control measures should be clear, concise, executable orders.

Note. Most vital to developing CRM control measures is mature, educated leadership.

Educational Control Measures

2-76. Educational control measures occur when adequate training takes place. They require the largest amount of planning and training time. Leaders implement educational control measures using two sequential steps:
    (1) Supervisors and instructors must be certified.
    (2) Soldier training must be executed.

Physical Control Measures

2-77. Physical control measures are the measures emplaced to reduce injuries. This includes not only protective equipment, but also certified personnel to supervise the training. Unrestrained physical control measures are, in themselves, a hazard.

IMPLEMENT CONTROL MEASURES

2-78. When leaders implement the control measures, they must match the control measures to the Soldier's skill level. They must also enforce every control measure as a means of validating its adequacy.

SUPERVISE AND EVALUATE

2-79. This step enables leaders to eliminate unnecessary risk and ineffective control measures by identifying unexpected hazards and determining if the implemented control measures reduced the residual risk without interfering with the training.

CONDUCT AN ENVIRONMENTAL RISK ASSESSMENT

2-80. All leaders, trainers, and Soldiers must comply with environmental laws and regulations. The leader must identify the environmental risks associated with training individual and collective tasks, and implement environmental protection measures by integrating them into plans, orders, SOPs, training performance standards, and rehearsals.

2-81. Environmental risk management parallels safety risk management and is based on the same philosophy. Environmental risk management consists of identifying hazards before they happen and assessing hazards caused during training.

Note. See FM 5-19 for more information.

IDENTIFY HAZARDS

2-82. Leaders should identify the potential sources for environmental degradation during the analysis of mission, enemy, terrain, troops, time available, and civil considerations (METT-TC). An environmental hazard is a condition with the potential for polluting air, soil, or water or destroying cultural or historical artifacts.

ASSESS HAZARDS

2-83. Leaders should analyze the potential severity of environmental degradation by using the environmental risk assessment matrixes in FM 5-19. The risk effect value is defined as an indicator of the severity of environmental degradation. Leaders quantify the environmental risk resulting from the operation as extremely high, medium, or low using the environmental assessment matrixes.

MAKE RANGE COORDINATIONS

2-84. Once the risk assessment is completed, viewed, and command approved, the OIC or NCOIC should check out the range and coordinate for range use.

> *Note.* The OIC or NCOIC should coordinate at least one day ahead of actual use to rehearse range setup and conduct.

TRAINING AIDS AND DEVICES

2-85. Appendix B of this manual outlines the training aids and devices used for shoulder-launched munition training.

RANGES

2-86. Shoulder-launched munition training requires a range complex that meets specific standards. This complex may be used for multilevel training and firing of shoulder-launched munitions.

> *Note.* See TC 25-8 for more information about ranges.

Authorized Ammunition

2-87. Subcaliber training launchers and live shoulder-launched munitions may be used on the same range. However, preliminary, basic, and advanced firing tables require moving target engagements. Most Army ranges that authorize both small arms ammunition and high-explosive (HE) rounds do not have or may not support the use of moving targets.

> *Note.* To learn more about range specifications, check post range regulations.

Training Areas

2-88. Training areas should be near, but not adjacent to, the firing line.

Firing Line

2-89. The firing line should be designed to allow personnel to observe firing from the side. Firing points should be positioned to allow 100 meters for backblast (Figure 2-17) and at least 20 meters between firing points.

> *Note.* The distance between firing points allows for a rear safe area when engaging moving targets.

Backblast Area

2-90. The backblast area must be fenced, roped, or marked in some way, so Soldiers know not to enter it when firing is being conducted.

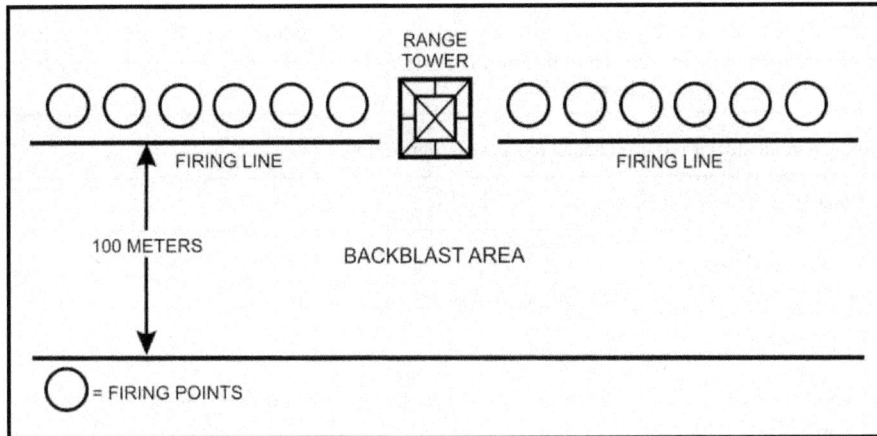

Figure 2-17. Firing line.

TARGET ARRAY

2-91. The target array should include stationary and moving vehicle targets and bunker targets at ranges of 100 to 300 meters. TADSS (Multiple Integrated Laser Engagement System [MILES], EST 2000) enable the unit to conduct practice fire on targets located beyond the munition's maximum effective range.

> *Note.* Firers may use live HE munitions to engage hard targets (armor vehicles) only. This reduces the damage to other targets the unit must maintain, such as multipurpose range complex (MPRC) target systems and bunkers.

Stationary Targets

2-92. Standard vehicle silhouettes or tank hulls should be used for stationary vehicle targets. Bunker silhouettes should have a dark, painted, 1-meter square at the bottom/center of a 4- by 8-foot plywood target or manmade structure.

2-93. Stationary targets should be positioned to—
- Allow firers to engage flank, frontal, and oblique targets.
- Accommodate a vehicle and bunker target array at ranges of 100 to 300 meters.

Moving Targets

2-94. Moving target silhouettes should travel along a track or road so the firers can engage fast-moving targets from both flank and oblique angles. Moving targets should be placed at ranges of 100 to 300 meters.

EQUIPMENT

2-95. The following is the minimum amount of range materiel and supplies needed to operate a practice- or live-fire shoulder-launched munition range:
- A helmet, a body armor vest, load-carrying equipment (LCE) or an enhanced tactical load-bearing vest (ETLBV), and ear protection for all range personnel and Soldiers attending training.
- Appropriate publications pertaining to training (GS TMs, FMs, TMs, ARs, SOPs).
- Range flag.
- Communications equipment.
- Targets in accordance with this manual.
- Shoulder-launched munitions (live/practice), as needed.
- TADSS, as needed.

Note. TADSS enable Soldiers to learn as much as they can about a munition before they attempt to fire the actual munition. Their use saves money and time, and prevents injuries. See Appendix B for more information about TADSS.

- Ambulance or required dedicated evacuation vehicle.

Note. The driver must have knowledge of the route to the hospital.

- Potable water.
- Scorecards in accordance with this manual.

PERSONNEL

2-96. In accordance with DA PAM 385-63, the following safety personnel are required for shoulder-launched munition training (Table 2-4):
- OIC.
- Range safety officer (RSO).

Note. OICs and RSOs involved in serious range incidents may lose their certification if determined to be in violation of AR 385-63 or DA PAM 385-63. While an incident is under investigation, their certificate may be suspended for as long as deemed necessary or revoked by the installation commander.

2-97. Safe and successful performance of training also requires experienced support personnel. Support personnel required for training include—
- Safety NCOs.
- Ammunition personnel.
- Tower operator.
- Guards, as required.
- Medical personnel.
- Truck driver, if applicable.

Table 2-4. Officer in charge and range safety officer requirements.

SYSTEMS	PERSONNEL REQUIREMENTS	
	OIC	RSO
Subcaliber training launchers	SFC	SSG
Live shoulder-launched munitions	SFC	SSG
LFXs, using organic weapons (squad through company, battery, and troop)	SFC	SSG
Combined arms live-fire exercises (CALFEXs) using outside fire support (section, platoon, squad, company, battery, troop, battalion, and squadron or larger)[2]	SFC	SSG

[1] When chemical, biological, radiological, nuclear (CBRN) training is being conducted, the OIC/RSO must be CBRN-qualified.
[2] The OIC will be a field-grade officer for battalion and larger-size units. For CALFEXs, the RSO will be of the ranks listed above based on the complexity of the exercise and number of participants (i.e., squad, section, platoon, company, troop, squadron, battalion, and larger).

Note. Ranks of other services, Army civilians, and contractors must be equivalent to U.S. Army ranks.

Officer in Charge and Noncommissioned Officer in Charge

2-98. The OIC must have satisfactorily completed a standard program of instruction in the duties of the OIC (developed by the unit to which he is assigned) and attended a range safety briefing conducted by the installation range control. The OIC or NCOIC must—

- Be knowledgeable in the munitions and subcaliber training launchers involved and the duties required.
- Be certified by the commander.

Note. The rank of the OIC is determined by unit policies and regulations.

2-99. Once selected by the commander, the OIC should select the right personnel to conduct the training. Next, he should appoint a NCOIC who has current experience in the use of shoulder-launched munitions. The OIC and NCOIC should—

- Select and brief range support personnel on expected duties.
- Certify selected range personnel on their range duties.

Note. Before conducting training, the OIC and NCOIC should review unit SOPs, AR 385-63, and DA PAM 385-63.

Range Safety Officer

2-100. The RSO should be the senior shoulder-launched munition instructor. The RSO must have satisfactorily completed a standard program of instruction in the duties of RSO (developed by the unit to which he is assigned) and attended a range safety briefing conducted by the installation range control. The RSO must—

- Be an E6 or above.
- Be knowledgeable in the munitions and subcaliber training launchers involved and the duties required.
- Ensure that the OIC has current safety cards.
- Perform no duties other than those of RSO.

Safety Noncommissioned Officers

2-101. Safety NCOs provide instruction, prepare shoulder-launched munitions, and conduct practice and live shoulder-launched munition training safely. Safety NCOs should—

- Be an E5 or above.
- Be knowledgeable in the munitions and subcaliber training launchers involved and the duties required.
- Be selected and certified on all shoulder-launched munition tasks by the OIC and NCOIC.
- Ensure that no firers are forward of or behind the person to his right or left on the firing line.
- Ensure that the firer identifies the correct target.
- Prepare and load all subcaliber training launchers.
- Ensure that the operator performs misfire procedures correctly, or correct the problem.
- Signal to the tower operator when firers are ready to fire.
- Move firers on and off the firing line.
- Provide shoulder-launched munition instruction.
- Demonstrate live-fire munitions.

Note. These personnel require no safety cards, but must be task-certified by their unit on all shoulder-launched munition tasks.

Ammunition Personnel

2-102. The ammunition personnel are in charge of accountability and handing out shoulder-launched munitions.

> *Note.* The ammunition NCO must attend an ammunition handler's class provided by the local ammunition supply point (ASP).

Tower Operator

2-103. The tower operator scores target hits, controls Soldier movements during range operations, and monitors communications with range control.

Guards

2-104. Guards control vehicle and foot traffic entering the range during range operations.

Medical Personnel

2-105. Medical support (with required medical supplies) must be present before and during range operations.

Truck Driver

2-106. The truck driver transports personnel to and from the range and provides support as needed (e.g., water, food, guard, etc.).

SECTION IV. TRAINING CONDUCT

2-107. Training conduct involves four steps:

    (1) Occupy, inspect, and set up the range.
    (2) Prepare for training.
    (3) Conduct the training.
    (4) Complete the training mission.

OCCUPY, INSPECT, AND SET UP RANGE

2-108. The OIC must establish communication with the installation's range control and request permission to occupy the range before personnel, materiel, or supplies arrive. Once this has been accomplished, the OIC and NCOIC should—

- Set up ammunition points and post guards.
- Establish locations for a medical station.
- Designate Soldier holding areas.
- Establish water points.
- Designate parking areas.
- Inspect the range for operational conditions.
- Request an opening code from range control, if applicable.
- Raise the range flag.

PREPARE FOR TRAINING

2-109. The OIC and NCOIC should greet unit leaders and Soldiers as they arrive and direct them to the holding area. Actions at the holding area include the following:

- Ensure all Soldiers attending training have a helmet, a body armor vest, LCE/ETLBV, ear protection, and a protective mask.
- Identify Soldiers to be trained.
- Conduct a safety briefing (to include administrative personnel).

CONDUCT THE TRAINING

Note. The OIC should monitor all training activities.

2-110. Soldiers progress through three phases of training:
- Preliminary.
- Basic.
- Advanced.

2-111. Two forms are used to record the results of this training:
- DA Form 7676 (Day and Night Fire—M141 BDM [BDM Subcaliber Training Launcher], Figure 2-18).
- DA Form 7677 (Day and Night Fire—M136 AT4 [M287 Subcaliber Training Launcher], Figure 2-19).

Note. Copies of these forms are located at the end of this publication for local reproduction on 8 1/2-by 11-inch paper.

WARNING

When firing the BDM subcaliber training launcher at targets less than 100 meters away, fire from behind safe cover (i.e., sandbag wall, concrete wall) to prevent injury from flying debris.

Do not fire the M287 subcaliber training launcher at target ranges of less than 125 meters due to ricochet.

Dangerous noise levels exist within 100 meters of a fired shoulder-launched munition. Operator and personnel must wear properly fitted, approved earplugs to prevent permanent hearing loss or damage.

PRELIMINARY

2-112. Preliminary shoulder-launched munition marksmanship training covers those tasks learned during a Soldier's initial training, as well as target engagement procedures. During this phase of training, Soldiers receive instruction and perform hands-on training using a FET or FHT. This instruction covers the following tasks:
- Perform serviceability checks on a shoulder-launched munition.
- Prepare a shoulder-launched munition for firing.
- Demonstrate correct firing positions.
- Estimate range to a target.
- Apply the fundamentals of marksmanship (to obtain a correct sight picture), such as—
  - Steady hold.
  - Aiming procedures, including eye placement, sight alignment, and sight picture.
  - Breath control.
  - Trigger manipulation.
- Perform misfire procedures.
- Return the shoulder-launched munition to the carrying configuration.

Notes. 1. Preliminary tasks should be taught or reinforced before conducting any form of live-fire training.

2. With proper training and oversight by the instructor/trainer, a Soldier with poor marksmanship skills can improve those skills with the help of the EST 2000. See Appendix B for more information about the EST 2000.

2-113. Trainers administer performance evaluations to determine how well Soldiers perform against established performance measures. Those who fail are retrained and retested, and those who pass help retrain and evaluate those who did not.

BASIC

2-114. Basic shoulder-launched munition marksmanship training begins at the conclusion of Phase I. During this phase of training, Soldiers receive instruction and perform hands-on training using a FHT, subcaliber training launchers, and other TADSS. This instruction covers the following tasks:

- Identify targets.
- Determine target engageability.
- Engage targets using MOPP engagement techniques (practice day fire).
- Install and operate NVSs.
- Install and operate aiming lights.
- Use other TADSS.

Note. For more information about TADSS, see Appendix B.

2-115. Trainers administer performance evaluations to determine how well Soldiers perform against established performance measures. Those who fail are retrained and retested, and those who pass help retrain and evaluate those who did not.

2-116. Following this training, Soldiers perform practice fire (using subcaliber training launchers) in daytime and limited visibility conditions.

Practice Day Fire

2-117. Soldiers conduct practice day fire using a BDM subcaliber training launcher and a M287 subcaliber training launcher.

Note. During the execution of this training, range safety personnel should load the subcaliber training launcher and perform any necessary maintenance.

2-118. Table 2-5 shows the distribution of rounds.

Table 2-5. Distribution of rounds for practice day fire.

ROUND	TYPE OF TARGET	RANGE (METERS)	FIRING POSITION
M141 BDM (BDM SUBCALIBER TRAINING LAUNCHER)			
1	Stationary	100 to 200	Standing
2	Stationary	100 to 250	Modified Kneeling*
3	Moving	100 to 250	Standing
M136 AT4 (M287 SUBCALIBER TRAINING LAUNCHER)			
1	Stationary	125 to 200	Standing
2	Stationary	125 to 300	Modified Kneeling*
3	Moving	125 to 250	Standing
* The firer wears MOPP gear when firing from this position.			

DAY AND NIGHT FIRE
M141 BDM (BDM SUBCALIBER TRAINING LAUNCHER)

For use of this form, see TM 3-23.25. The proponent agency is TRADOC.

NAME _John Doe_		RANK _SFC_	UNIT _B Co. 2nd BN 29th Reg7_
DATE _4/22/2010_	EVALUATOR'S NAME _James Smith_		EVALUATOR'S RANK _SFC_

TABLE 1—PRACTICE DAY FIRE

ROUND	TYPE OF TARGET	RANGE (METERS)	FIRING POSITION	HIT	MISS
1	Stationary	100 to 200	Standing	X	
2	Stationary	100 to 250	Modified Kneeling*	X	
3	Moving	100 to 250	Standing	X	
TOTAL				3	0

* The firer wears MOPP gear when firing from this position.

TABLE 2— PRACTICE NIGHT FIRE

ROUND	TYPE OF TARGET	RANGE (METERS)	FIRING POSITION	HIT	MISS
1	Stationary	100 to 250	Standing	X	
2	Moving	100 to 200	Kneeling		X
TOTAL				1	1

TABLE 3—QUALIFICATION DAY FIRE

ROUND	TYPE OF TARGET	RANGE (METERS)	FIRING POSITION	HIT	MISS
1	Stationary	100 to 200	Standing	X	
2	Stationary	100 to 250	Modified Kneeling	X	
3	Moving	100 to 250	Kneeling	X	
4	Stationary	100 to 200	Modified Kneeling	X	
5	Stationary	100 to 250	Standing	X	
6	Moving	100 to 200	Standing	X	
TOTAL				6	0

TABLE 4—QUALIFICATION NIGHT FIRE

ROUND	TYPE OF TARGET	RANGE (METERS)	FIRING POSITION	HIT	MISS
1	Stationary	100 to 250	Standing	X	
2	Stationary	100 to 250	Modified Kneeling	X	
3	Moving	100 to 250	Kneeling		X
TOTAL				2	

PRACTICE FIRE SCORE			QUALIFICATION FIRE SCORE		
TABLE	HIT	MISS	TABLE	HIT	MISS
1	3		3	6	0
2	1	1	4	2	0

THE FIRER WILL BE ISSUED 14 ROUNDS: 5 ROUNDS FOR THE PRACTICE FIRE TABLES AND 9 ROUNDS FOR THE QUALIFICATION FIRE TABLES.

AT RANGES BEYOND 250 METERS, SOLDIERS WILL NOT BE ABLE TO OBSERVE TRACER IMPACT. THIS ISSUE IS RESOLVED WHEN USING AN MPRC, AS TARGETS MOVE DOWN UPON IMPACT.

QUALIFICATION SCORE RATINGS

- ☐ 9 OF 9 HITS — EXPERT
- ☒ 8 OF 9 HITS — 1st CLASS
- ☐ 6 OF 9 HITS — 2nd CLASS
- ☐ 5 AND BELOW — UNQUALIFIED

FIRER'S SIGNATURE _John Doe_	DATE _4/22/2010_	SCORER'S INITIALS _JS_	DATE INITIALED _4/22/2010_
RANGE OIC'S SIGNATURE _Henry Wone_		RANK _SFC_	DATE _4/22/2010_

DA Form 7676, OCT 2010

APD PE v1.00ES

Figure 2-18. Example of completed DA Form 7676
(Day and Night Fire—M141 BDM [BDM Subcaliber Training Launcher]).

DAY AND NIGHT FIRE
M136 AT4 (M287 SUBCALIBER TRAINING LAUNCHER)

For use of this form, see TM 3-23.25. The proponent agency is TRADOC.

NAME John Doe	RANK SPC	UNIT B Co 2nd BN/29th Regt
DATE 22 April 2010	EVALUATOR'S NAME James Smith	EVALUATOR'S RANK SGT

TABLE 1—PRACTICE DAY FIRE

ROUND	TYPE OF TARGET	RANGE (METERS)	FIRING POSITION	HIT	MISS
1	Stationary	125 to 200	Standing	X	
2	Stationary	125 to 300	Modified Kneeling*	X	
3	Moving	125 to 250	Standing		X
TOTAL				2	1

* The firer wears MOPP gear when firing from this position.

TABLE 2— PRACTICE NIGHT FIRE

ROUND	TYPE OF TARGET	RANGE (METERS)	FIRING POSITION	HIT	MISS
1	Stationary	125 to 250	Standing	X	
2	Moving	125 to 200	Kneeling		X
TOTAL				1	1

TABLE 3—QUALIFICATION DAY FIRE

ROUND	TYPE OF TARGET	RANGE (METERS)	FIRING POSITION	HIT	MISS
1	Stationary	125 to 200	Standing	X	
2	Stationary	125 to 300	Modified Kneeling	X	
3	Moving	125 to 250	Kneeling		X
4	Stationary	125 to 200	Modified Kneeling	X	
5	Stationary	125 to 300	Standing	X	
6	Moving	125 to 250	Standing		X
TOTAL				4	2

TABLE 4—QUALIFICATION NIGHT FIRE

ROUND	TYPE OF TARGET	RANGE (METERS)	FIRING POSITION	HIT	MISS
1	Stationary	125 to 250	Standing	X	
2	Stationary	125 to 300	Modified Kneeling	X	
3	Moving	125 to 250	Kneeling		X
TOTAL				2	1

PRACTICE FIRE SCORE			QUALIFICATION FIRE SCORE		
TABLE	HIT	MISS	TABLE	HIT	MISS
1	2	1	3	4	2
2	1	1	4	2	1

THE FIRER WILL BE ISSUED 14 ROUNDS: 5 ROUNDS FOR THE PRACTICE FIRE TABLES AND 9 ROUNDS FOR THE QUALIFICATION FIRE TABLES.

AT RANGES BEYOND 550 METERS, SOLDIERS WILL NOT BE ABLE TO OBSERVE TRACER IMPACT. THIS ISSUE IS RESOLVED WHEN USING AN MPRC, AS TARGETS MOVE DOWN UPON IMPACT.

QUALIFICATION SCORE RATINGS

☐ 9 OF 9 HITS — EXPERT

☐ 8 OF 9 HITS — 1st CLASS

☒ 6 OF 9 HITS — 2nd CLASS

☐ 5 AND BELOW — UNQUALIFIED

FIRER'S SIGNATURE John Doe	DATE 4/22/2010	SCORER'S INITIALS JS	DATE INITIALED 4/22/2010
RANGE OIC'S SIGNATURE Henry Moore		RANK SFC	DATE 4/22/2010

DA Form 7677, OCT 2010   APD PE v1.00ES

**Figure 2-19. Example of completed DA Form 7677
(Day and Night Fire—M136 AT4 [M287 Subcaliber Training Launcher]).**

M141 Bunker Defeat Munition (Bunker Defeat Munition Subcaliber Training Launcher)

2-119. Soldiers fire three rounds using a BDM subcaliber training launcher: two rounds at bunker targets at ranges of 100 to 250 meters and one round at a moving target at a range of 100 to 250 meters. The purpose of this firing is to determine the firer's ability to estimate range to the target during day conditions, demonstrate correct firing positions, apply the fundamentals of marksmanship, and achieve accuracy while receiving blast overpressure effects. Table 2-6 shows the task, conditions, and standards for this training.

> *Note.* Soldier accuracy deteriorates after experiencing the blast effects of the initial round. Firing assessments prove that blast anticipation after firing the initial round causes the firer to concentrate more on blast effects than the target. This can be overcome if Soldiers are given the opportunity to fire more shoulder-launched munitions and at a greater frequency. Soldiers can use simulators that closely replicate the blast effects of firing live munitions to reduce firer anticipation.

Table 2-6. Task, conditions, and standards for practice day fire (bunker defeat munition subcaliber training launcher).

TASK	Engage a target with a BDM subcaliber training launcher.
CONDITIONS	On a suitable MPRC. Given one BDM subcaliber training launcher and three HA21 training rockets, two bunker targets at ranges of 100 to 250 meters, and one target moving 8 to 24 km per hour at a range of 100 to 250 meters.
STANDARD	The Soldier fires three rockets at stationary and moving targets and achieves at least two hits. The Soldier demonstrates correct firing positions, estimates range to the target, and applies the fundamentals of marksmanship.

2-120. The results are recorded on Table 1 of DA Form 7676.

M136 AT4 (M287 Subcaliber Training Launcher)

2-121. Soldiers fire three rounds using a M287 subcaliber training launcher: two rounds at stationary targets at ranges of 125 to 300 meters and one round at a moving target at a range of 125 to 250 meters. The purpose of this firing is to determine the firer's ability to estimate range to the target during day conditions, demonstrate correct firing positions, apply the fundamentals of marksmanship, and achieve accuracy. Table 2-7 shows the task, conditions, and standards for this training.

Table 2-7. Task, conditions, and standards for practice day fire (M287 subcaliber training launcher).

TASK	Engage a target with a M287 subcaliber training launcher.
CONDITIONS	On a suitable MPRC. Given one M287 subcaliber training launcher and three rounds of M939 9-mm training practice-tracer (TP-T) ammunition, two stationary targets at ranges of 125 to 300 meters, and one target moving 8 to 24 km per hour at a range of 125 to 250 meters.
STANDARD	The Soldier fires three tracer bullets at stationary and moving targets and achieves at least two hits. The Soldier demonstrates correct firing positions, estimates range to the target, and applies the fundamentals of marksmanship.

2-122. The results are recorded on Table 1 of DA Form 7677.

Practice Night Fire

2-123. Soldiers conduct practice night fire using a BDM subcaliber training launcher and a M287 subcaliber training launcher.

> *Notes.* 1. During the execution of this training, range safety personnel should load the subcaliber training launcher and perform any necessary maintenance.
>
> 2. Practice night fire consists of hands-on installation of NVDs and firing. Instructors will prepare all shoulder-launched munitions for conducting night fire.

2-124. Table 2-8 shows the distribution of rounds.

Table 2-8. Distribution of rounds for practice night fire.

ROUND	TYPE OF TARGET	RANGE (METERS)	FIRING POSITION
M141 BDM (BDM SUBCALIBER TRAINING LAUNCHER)			
1	Stationary	100 to 250	Standing
2	Moving	100 to 200	Kneeling
M136 AT4 (M287 SUBCALIBER TRAINING LAUNCHER)			
1	Stationary	125 to 250	Standing
2	Moving	125 to 200	Kneeling

M141 Bunker Defeat Munition (Bunker Defeat Munition Subcaliber Training Launcher)

2-125. Soldiers fire two rounds using a BDM subcaliber training launcher: one round at a bunker target at a range of 100 to 250 meters and one round at a moving target at a range of 100 to 200 meters. The purpose of this firing is to determine the firer's ability to estimate range to the target during limited visibility conditions, demonstrate correct firing positions, apply the fundamentals of marksmanship, and achieve accuracy while receiving blast overpressure effects. Table 2-9 show the task, conditions, and standards for this training.

> *Note.* Soldier accuracy deteriorates after experiencing the blast effects of the initial round. Firing assessments prove that blast anticipation after firing the initial round causes the firer to concentrate more on blast effects than the target. This can be overcome if Soldiers are given the opportunity to fire more shoulder-launched munitions and at a greater frequency. Soldiers can use simulators that closely replicate the blast effects of firing live munitions to reduce firer anticipation.

**Table 2-9. Task, conditions, and standards for practice night fire
(bunker defeat munition subcaliber training launcher).**

TASK	Engage a target with a BDM subcaliber training launcher.
CONDITIONS	On a suitable MPRC. Given one BDM subcaliber training launcher, two HA21 training rockets, mounted NVDs, one bunker target at a range of 100 to 250 meters, and one target moving 8 to 24 km per hour at a range of 100 to 200 meters.
STANDARD	The Soldier fires two rockets at stationary and moving targets and achieves at least one hit. The Soldier demonstrates correct firing positions, estimates range to the target, and applies the fundamentals of marksmanship.

2-126. The results are recorded on Table 2 of DA Form 7676.

M136 AT4 (M287 Subcaliber Training Launcher)

2-127. Soldiers fire two rounds using a M287 subcaliber training launcher: one round at a stationary target at a range of 125 to 250 meters and one round at a moving target at a range of 125 to 200 meters. The purpose of this firing is to determine the firer's ability to estimate range to the target during limited visibility conditions, demonstrate correct firing positions, apply the fundamentals of marksmanship, and achieve accuracy. Table 2-10 shows the task, conditions, and standards for this training.

**Table 2-10. Task, conditions, and standards for practice night fire
(M287 subcaliber training launcher).**

TASK	Engage a target with a M287 subcaliber training launcher.
CONDITIONS	On a suitable MPRC. Given one M287 subcaliber training launcher and two rounds of M939 9-mm TP-T ammunition, mounted NVDs, one stationary target at a range of 125 to 250 meters, and one target moving 8 to 24 km per hour at a range of 125 to 200 meters.
STANDARD	The Soldier fires two tracer bullets at stationary and moving targets and achieves at least one hit. The Soldier demonstrates correct firing positions, estimates range to the target, and applies the fundamentals of marksmanship.

2-128. The results are recorded on Table 2 of DA Form 7677.

Advanced

2-129. Advanced shoulder-launched munition marksmanship training begins at the conclusion of Phase II. During advanced marksmanship training, Soldiers receive instruction and hands-on training on the following tasks:

- Load and maintain the subcaliber training launchers.
- Use and maintain other TADSS.
- Use the proper methods of engagement (single, sequence, pair, and volley fire).
- Use and maintain NVSs, and perform sight alignment procedures.
- Use and maintain aiming lights, and perform sight alignment procedures.

2-130. Trainers administer performance evaluations to determine how well Soldiers perform against established performance measures. Those who fail are retrained and retested, and those who pass help retrain and evaluate those who did not.

2-131. Following this training, qualification day and night fire are conducted.

Qualification Day Fire

2-132. Soldiers conduct qualification day fire using a BDM subcaliber training launcher and a M287 subcaliber training launcher. Table 2-11 shows the distribution of rounds.

Table 2-11. Distribution of rounds for qualification day fire.

ROUND	TYPE OF TARGET	RANGE (METERS)	FIRING POSITION
M141 BDM (BDM SUBCALIBER TRAINING LAUNCHER)			
1	Stationary	100 to 200	Standing
2	Stationary	100 to 250	Modified Kneeling
3	Moving	100 to 250	Kneeling
4	Stationary	100 to 200	Modified Kneeling
5	Stationary	100 to 250	Standing
6	Moving	100 to 200	Standing
M136 AT4 (M287 SUBCALIBER TRAINING LAUNCHER)			
1	Stationary	125 to 200	Standing
2	Stationary	125 to 300	Modified Kneeling
3	Moving	125 to 250	Kneeling
4	Stationary	125 to 200	Modified Kneeling
5	Stationary	125 to 300	Standing
6	Moving	125 to 200	Standing

M141 Bunker Defeat Munition (Bunker Defeat Munition Subcaliber Training Launcher)

2-133. Soldiers fire six rounds using a BDM subcaliber training launcher: four rounds at bunker targets at ranges of 100 to 250 meters and two rounds at moving targets at ranges of 100 to 250 meters. The purpose of this firing is to determine the firer's ability to estimate range to the target during day conditions, demonstrate correct firing positions, apply the fundamentals of marksmanship, and achieve accuracy while receiving blast overpressure effects. Table 2-12 shows the task, conditions, and standards for this training.

Notes. 1. During the execution of this training, Soldiers should load the subcaliber training launcher and perform any necessary maintenance.

2. Soldier accuracy deteriorates after experiencing the blast effects of the initial round. Firing assessments prove that blast anticipation after firing the initial round causes the firer to concentrate more on blast effects than the target. This can be overcome if Soldiers are given the opportunity to fire more shoulder-launched munitions and at a greater frequency. Soldiers can use simulators that closely replicate the blast effects of firing live munitions to reduce firer anticipation.

2-134. The results are recorded on Table 3 of DA Form 7676.

Table 2-12. Task, conditions, and standards for qualification day fire (bunker defeat munition subcaliber training launcher).

TASK	Engage a target with a BDM subcaliber training launcher.
CONDITIONS	On a suitable MPRC. Given one BDM subcaliber training launcher and six HA21 training rockets, four bunker targets at ranges of 100 to 250 meters, and two targets moving 8 to 24 km per hour at ranges of 100 to 250 meters.
STANDARD	The Soldier fires six rockets at stationary and moving targets and achieves at least four hits. The Soldier demonstrates correct firing positions, estimates range to the target, and applies the fundamentals of marksmanship.

M136 AT4 (M287 Subcaliber Training Launcher)

2-135. Soldiers fire six rounds using a M287 subcaliber training launcher: four rounds at stationary targets at ranges of 125 to 300 meters and two rounds at moving targets at ranges of 125 to 250 meters. The purpose of this firing is to determine the firer's ability to estimate range to the target during day conditions, demonstrate correct firing positions, apply the fundamentals of marksmanship, and achieve accuracy. Table 2-13 shows the task, conditions, and standards for this training.

> *Note.* During the execution of this training, Soldiers should load the subcaliber training launcher and perform any necessary maintenance.

Table 2-13. Task, conditions, and standards for qualification day fire (M287 subcaliber training launcher).

TASK	Engage a target with a M287 subcaliber training launcher.
CONDITIONS	On a suitable MPRC. Given one M287 subcaliber training launcher and six rounds of M939 9-mm TP-T ammunition, four stationary targets at ranges of 125 to 300 meters, and two targets moving 8 to 24 km per hour at ranges of 125 to 250 meters.
STANDARD	The Soldier fires six tracer bullets at stationary and moving targets and achieves at least four hits. The Soldier demonstrates correct firing positions, estimates range to the target, and applies the fundamentals of marksmanship.

2-136. The results are recorded on Table 3 of DA Form 7677.

Qualification Night Fire

2-137. Soldiers conduct qualification night fire using a BDM subcaliber training launcher and a M287 subcaliber training launcher. Table 2-14 shows the distribution of rounds.

Table 2-14. Distribution of rounds for qualification night fire.

ROUND	TYPE OF TARGET	RANGE (METERS)	FIRING POSITION
M141 BDM (BDM SUBCALIBER TRAINING LAUNCHER)			
1	Stationary	100 to 250	Standing
2	Stationary	100 to 250	Modified Kneeling
3	Moving	100 to 250	Kneeling
M136 AT4 (M287 SUBCALIBER TRAINING LAUNCHER)			
1	Stationary	125 to 250	Standing
2	Stationary	125 to 300	Modified Kneeling
3	Moving	125 to 250	Kneeling

M141 Bunker Defeat Munition (Bunker Defeat Munition Subcaliber Training Launcher)

2-138. Soldiers fire three rounds using a BDM subcaliber training launcher: two rounds at bunker targets at ranges of 100 to 250 meters and one round at a moving target at a range of 100 to 250 meters. The purpose of this firing is to determine the firer's ability to estimate range to the target during limited visibility conditions, demonstrate correct firing positions, apply the fundamentals of marksmanship, and achieve accuracy while receiving blast overpressure effects. Table 2-15 shows the task, conditions, and standards for this training.

Notes. 1. During the execution of this training, Soldiers should load the subcaliber training launcher and perform any necessary maintenance.

2. Soldier accuracy deteriorates after experiencing the blast effects of the initial round. Firing assessments prove that blast anticipation after firing the initial round causes the firer to concentrate more on blast effects than the target. This can be overcome if Soldiers are given the opportunity to fire more shoulder-launched munitions and at a greater frequency. Soldiers can use simulators that closely replicate the blast effects of firing live munitions to reduce firer anticipation.

2-139. The results are recorded on Table 4 of DA Form 7676.

Table 2-15. Task, conditions, and standards for qualification night fire (bunker defeat munition subcaliber training launcher).

TASK	Engage a target with a BDM subcaliber training launcher.
CONDITIONS	On a suitable MPRC. Given one BDM subcaliber training launcher and three HA21 training rockets, mounted NVDs, and two bunker targets at ranges of 100 to 250 meters, and one target moving 8 to 24 km per hour at a range of 100 to 250 meters.
STANDARD	The Soldier fires three rockets at stationary and moving targets and achieves at least two hits. The Soldier demonstrates correct firing positions, estimates range to the target, and applies the fundamentals of marksmanship.

M136 AT4 (M287 Subcaliber Training Launcher)

2-140. During qualification night fire, Soldiers fire three rounds using a M287 subcaliber training launcher: two rounds at stationary targets at ranges of 125 to 300 meters and one round at a moving target at a range of 125 to 250 meters. The purpose of this firing is to determine the firer's ability to estimate range to the target during limited visibility conditions, demonstrate correct firing positions, apply the fundamentals of marksmanship, and achieve accuracy. Table 2-16 shows the task, conditions, and standards for this training.

Note. During the execution of this training, Soldiers should load the subcaliber training launcher and perform any necessary maintenance.

Table 2-16. Task, conditions, and standards for qualification night fire (M287 subcaliber training launcher).

TASK	Engage a target with a M287 subcaliber training launcher.
CONDITIONS	On a suitable MPRC. Given one M287 subcaliber training launcher and three rounds of M939 9-mm TP-T ammunition, mounted NVDs, two stationary targets at ranges of 125 to 300 meters, and one target moving 8 to 24 km per hour at a range of 125 to 250 meters.
STANDARD	The Soldier fires three tracer bullets at stationary and moving targets and achieves at least two hits. The Soldier demonstrates correct firing positions, estimates range to the target, and applies the fundamentals of marksmanship.

2-141. The results are recorded on Table 4 of DA Form 7677.

COLLECTIVE

2-142. Individual tasks must be integrated into collective training and rehearsals. To accomplish this, commanders analyze the collective tasks from their unit METLs and the individual tasks that support these collective tasks.

Collective Tasks

2-143. Table 2-17 contains a sample of collective tasks that shoulder-launched munitions may support.

Note. For more information about these collective tasks, see the Reimer Digital Library.

Individual Tasks Supported with Shoulder-Launched Munitions

2-144. Table 2-18 is a sample of individual tasks that shoulder-launched munitions may support.

Note. See STP 21-24-SMCT for more information.

Individual Shoulder-Launched Munitions Tasks

2-145. Table 2-19 is a task list of individual shoulder-launched munition tasks.

Note. See STP 21-1- SMCT for more information.

Table 2-17. Collective tasks that shoulder-launched munitions may support.

TASK NUMBER	TASK TITLE	SIZE OF ELEMENT
07-2-9001	Conduct an attack.	Company/Platoon
07-2-9003	Conduct a defense.	Company/Platoon
07-2-9004	Conduct a delay.	Company/Platoon
07-2-9008	Conduct a raid.	Company/Platoon
07-2-9009	Conduct a withdrawal.	Company/Platoon
07-2-9010	Conduct an ambush.	Company/Platoon
07-2-9011	Conduct tactical movement in an urban area.	Company/Platoon
07-3-9013	Conduct action on contact.	Platoon/Squad
07-3-9018*	Clear a building.	Platoon/Squad
07-3-9021*	Clear a trench line.	Platoon/Squad
Note. Tasks with an asterisk (*) are METT-TC dependent.		

Table 2-18. Sample of individual tasks that shoulder-launched munitions may support.

TASK NUMBER	TASK TITLE	SKILL LEVEL
071-410-0019	Control organic fires.	Levels 2, 3, and 4
071-420-0021	Conduct a movement to contact by a platoon.	Level 3
191-377-4203*	Supervise the establishment of a roadblock/checkpoint.	Level 3
191-378-5315*	Supervise an installation access control point.	Level 3
191-379-4407*	Plan convoy security operations.	Level 3
071-326-5805*	Conduct a route reconnaissance.	Level 4
071-430-0006	Conduct a defense by a platoon.	Level 4
551-721-4326*	Perform duties as convoy commander.	Level 4
181-101-4001*	Conduct a search and seizure.	Level 4
Notes.		
1. Skill Level 2 (E-5 team leader), Skill Level 3 (E-6 squad leader), Skill Level 4 (E-7 platoon leader).		
2. Tasks with an asterisk (*) are METT-TC dependent.		

Shoulder-Launched Munitions Hands-on Training Tasks

2-146. Table 2-20 is a task list of shoulder-launched munition hands-on training tasks.

Note. Soldiers must receive instruction and pass performance testing criteria before conducting LFXs.

Engagement Skills Trainer 2000

2-147. When planning training, unit trainers should incorporate shoulder-launched munitions training on the EST 2000 into the unit's training. EST 2000 provides both individual and collective scenarios for controlling organic fires, and allows the unit trainer to properly evaluate the task being conducted. The unit trainer must also identify deficiencies and correct them by retraining Soldiers until they can employ the selected munitions correctly.

Note. See Appendix B for more information about the EST 2000.

Force-on-Force Training

2-148. The MILES is a force-on-force trainer for shoulder-launched munitions. MILES is primarily used for force-on-force training; however, Soldiers can fire practice tables using MILES on a MPRC.

Notes.
1. The MILES should not be used for sustainment training.

2. Refer to DA PAM 350-38 for frequency of collective training.

3. See Appendix C for more information about the MILES.

COMPLETE INDIVIDUAL TRAINING MISSION

2-149. At the completion of training, all equipment, range materiel, and ammunition should be accounted for, range maintenance should be completed, and the OIC and RSO should close the range. This includes the performing the following tasks:
- Request a closing code from range control.
- Release unit Soldiers.
- Remove all equipment and ammunition from the range.

Note. Turn in all unexpended munitions to the ASP.

- Have the explosive ordnance disposal (EOD) unit find and clear any duds.
- Police the range, and perform other range maintenance as required by local SOP.
- Request a range inspection from range control when ready to clear.
- Turn in paperwork and equipment.
- Submit an after-action report to headquarters.
- Report any noted safety hazards to proper authorities.

Table 2-19. Shoulder-launched munition individual tasks.

TASK NUMBER	TASK TITLE	SKILL LEVEL
	M136 AT4	
071-054-0001	Prepare a M136 AT4 launcher for firing.	Level 1
071-054-0002	Restore a M136 AT4 launcher to carrying configuration.	Level 1
071-054-0003	Perform misfire procedures on a M136 AT4 launcher.	Level 1
071-054-0004*	Engage targets with a M136 AT4 launcher.	Level 1
	M136A1 AT4CS	
071-054-0011	Prepare a M136A1 AT4CS launcher for firing.	Level 1
071-054-0012	Restore a M136A1 AT4CS launcher to carrying configuration.	Level 1
071-054-0013	Perform misfire procedures on a M136A1 AT4CS launcher.	Level 1
071-054-0014*	Engage targets with a M136A1 AT4CS launcher.	Level 1
	M141 BDM	
071-054-0021	Prepare a M141BDM launcher for firing.	Level 1
071-054-0022	Restore a M141BDM launcher to carrying configuration.	Level 1
071-054-0023	Perform misfire procedures on a M141BDM launcher.	Level 1
071-054-0024*	Engage targets with a M141BDM launcher.	Level 1
Notes. 1. Skill Level 1 (E-1 thru E-4). 2. Tasks with an asterisk (*) are METT-TC dependent.		

Table 2-20. Hands-on training tasks.

REFERENCE	TASK
	M136 AT4—M136A1 AT4—M141 BDM
	Preliminary Training
TM 9-1340-228-10 BDM TM 9-1315-886-12 AT4 TM 9-1315-255-13 AT4CS	Perform serviceability checks on a shoulder-launched munition.
TM 9-1340-228-10 BDM TM 9-1315-886-12 AT4 TM 9-1315-255-13 AT4CS	Prepare a shoulder-launched munition for firing.
GS TM 3-23.25	Demonstrate correct firing positions.
	Identify targets.
	Estimate range to a target.
	Apply the fundamentals of marksmanship.
TM 9-1340-228-10 BDM TM 9-1315-886-12 AT4 TM 9-1315-255-13 AT4CS	Perform misfire procedures on a shoulder-launched munition.
TM 9-1340-228-10 BDM TM 9-1315-886-12 AT4 TM 9-1315-255-13 AT4CS	Restore a shoulder-launched munition to carrying configuration.
	Basic Training
GS TM 3-23.25	Determine target engageability.
	Select the appropriate shoulder-launched munition for a given target.
	Use other TADSS.
	Advanced Training
TM 9-1055-886-12&P M287	Load and maintain a subcaliber training launcher.
GS TM 3-23.25	Use and maintain other TADSS.
	Use proper methods of engaging targets (single, sequence, pair, and volley fire).
	Conduct sight alignment.
	Detect targets (stationary and moving).
	Determine target engageability.
	Select proper firing positions to engage targets.
Notes. 1. Skill Level 1 (E-1 through E-4). 2. Tasks with an asterisk (*) are METT-TC dependent.	

Chapter 3

PROCESS OF FIRING

The process of firing involves inspecting, preparing, and firing the munition. Further, Soldiers must understand proper care, handling, destruction, and decontamination procedures in order to effectively employ shoulder-launched munitions.

SECTION I. PACKAGING AND INSPECTION

3-1. Inspection involves examining the munition's packaging and the munition itself. All munitions should be inspected upon receipt.

M141 BUNKER DEFEAT MUNITION

3-2. M141 BDMs will usually be delivered to the supply point on pallets. This munition comes encased in a folding unit pack, which is secured in a metal ammunition container (Figure 3-1).

PACKAGING

3-3. The unit may receive M141 BDMs loaded on pallets and issue the munitions while held in their individual metal containers.

Note. The munitions should remain in the shipping container until Soldiers are ready to use them.

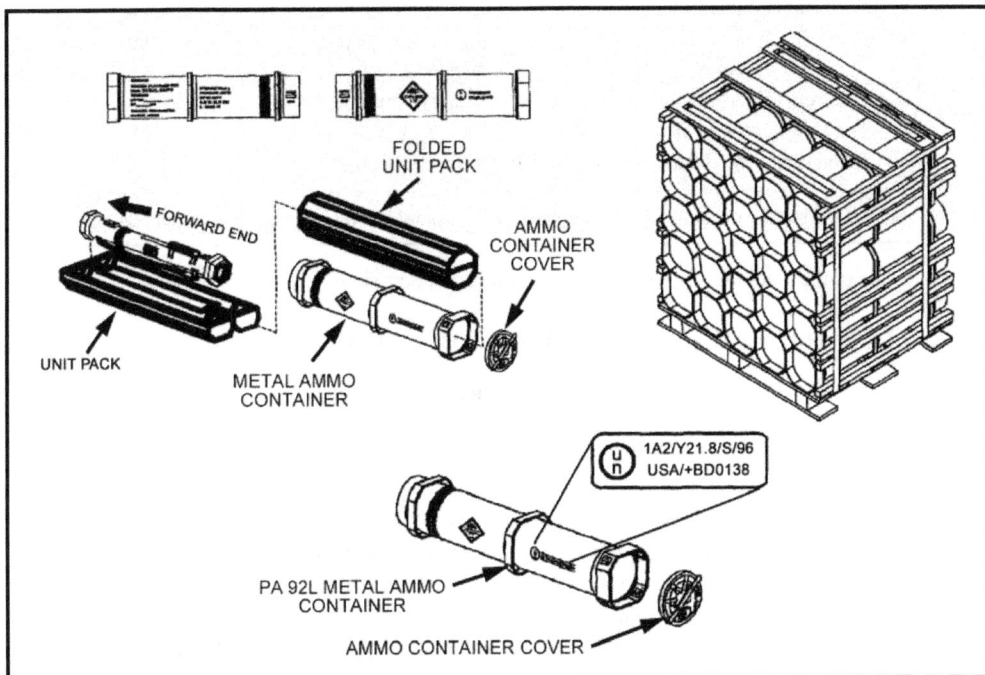

Figure 3-1. M141 bunker defeat munition packaging and pallet details.

3-4. Upon removal from the pallet, the metal ammunition containers should be individually inspected for damage. During this inspection, the containers must meet the following criteria:

- The container is not punctured or damaged.
- The container has the correct markings, including the serial number and bar code on the end of container.

3-5. If any of the containers do not meet these criteria, Soldiers should notify a supervisor.

MUNITION

3-6. Once a container passes inspection, it should be opened and the munition unpacked. To unpack the M141 BDM—

(1) Remove the metal ammunition container cover.

(2) Remove the unit pack from the container.

(3) Open the unit pack.

(4) Remove the munition.

3-7. Then, the Soldier should be able to observe and inspect the M141 BDM. The inspection is limited to a visual inspection of the outer tube and its components. During this inspection, the M141 BDM should meet the following criteria (Figure 3-2):

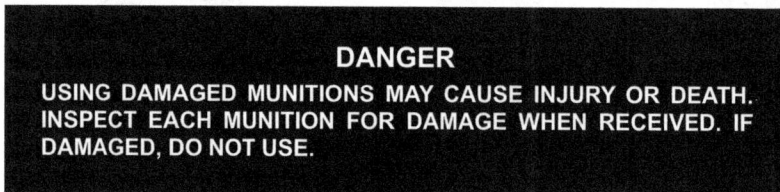

DANGER
USING DAMAGED MUNITIONS MAY CAUSE INJURY OR DEATH. INSPECT EACH MUNITION FOR DAMAGE WHEN RECEIVED. IF DAMAGED, DO NOT USE.

- The body is free of dents, cracks, gouges, or holes.
- The front and rear bumpers are present, and there are no holes, tears, or punctures.
- The front sight cover, rear sight cover, firing mechanism cover, tube release button cover, and shoulder stop cover are present and not damaged.
- The front and rear sights and the shoulder strap are present and not damaged.
- The sling shows no signs of fraying.
- When shaken, no sounds of loose or broken material come from inside the tube.

Note. If the munition is not to be used immediately, it should be returned to its shipping container.

3-8. If any of the munitions do not meet these criteria, Soldiers should notify a supervisor.

M136 AT4 SHOULDER-LAUNCHED MUNITION

3-9. M136 AT4s will usually be delivered to the supply point on pallets.

PACKAGING

3-10. Five M136 AT4s, each wrapped in a plastic barrier bag, are packed together in a wooden container. The containers are too heavy to stack more than four deep on the pallets (Figure 3-3).

3-11. Upon removal from the pallet, the wooden containers should be individually inspected for damage. During this inspection, the containers should meet the following criteria:

- The container is not punctured or damaged.
- The container has the correct markings, including the serial number and bar code on the end of the container.
- Fasteners are intact and show no signs of tampering.

3-12. If any of the containers do not meet these criteria, Soldiers should notify a supervisor.

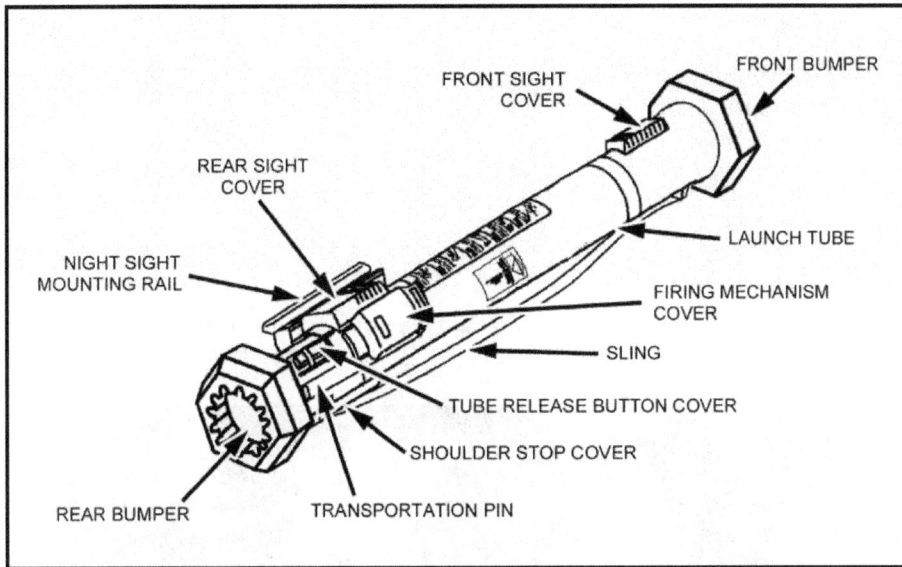

Figure 3-2. M141 bunker defeat munition inspection points.

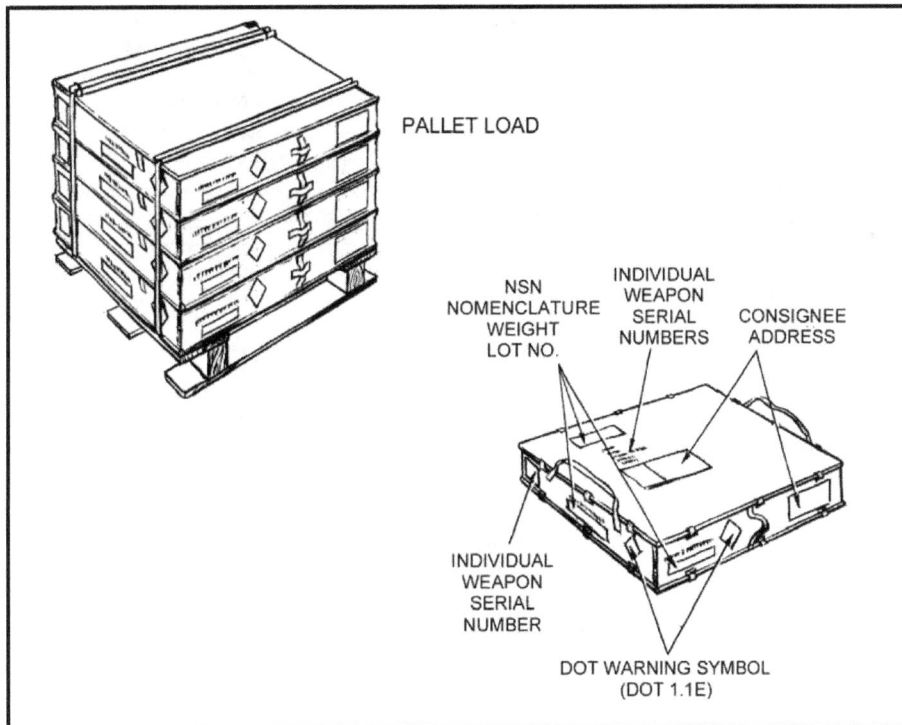

Figure 3-3. M136 AT4 packaging and pallet details.

MUNITION

3-13. Once a container passes inspection, it should be opened and the munition unpacked. To unpack the M136 AT4—

(1) Open the wooden container.

(2) Remove the plastic bags from the wooden container.

(3) Break the seal on the plastic bag.

Note. The plastic barrier bags have a V-notch cut 1 inch from the side to allow easy opening without the aid of a tool.

---

(4) Remove the munition.

3-14. The Soldier should then be able to observe and inspect the M136 AT4. The inspection is limited to a visual inspection of the outer tube and its components. During this inspection, the M136 AT4 should meet the following criteria (Figure 3-4):

> ## DANGER
>
> USING DAMAGED MUNITIONS MAY CAUSE INJURY OR DEATH. INSPECT EACH MUNITION FOR DAMAGE WHEN RECEIVED. IF DAMAGED, DO NOT USE.

- The rear seal, a brown acrylic plastic plate inside the venturi (located on the rear end of the munition), is in place and undamaged.
- The transport safety pin is in place and fully inserted. The lanyard is attached to the transport safety pin and the launcher. The lanyard should already be wrapped around the launcher clockwise, and the transport safety pin should be inserted in the retainer hole counterclockwise.
- The cocking lever is present and in the SAFE (uncocked) position.
- The plastic fire-through muzzle cover is in place and undamaged. If it is torn or broken, cut it out and check the launch tube to ensure it is clear of foreign objects. Remove any that you find by turning the tube muzzle downward and gently shaking the launcher.

Note. M136 AT4 launchers with missing muzzle covers and no obstruction are suitable for use. However, these launchers should have their muzzle covers replaced as soon as possible to prevent further damage and deterioration.

- The launcher has the correct color-coded band.
- The sights function properly. Open the sight covers to ensure the sights pop up and are undamaged.
- The red safety release catch does not move when you depress it.
- The red trigger button is not missing.
- The launcher body has no cracks, dents, or bulges.
- The carrying sling is not frayed and is attached firmly to the launch tube.
- The shoulder stop is not broken or damaged, and it unsnaps and folds down.

Note. If the M136 AT4 is not to be used immediately, it should be returned to its plastic bag and the bag resealed with tape.

3-15. If any of the munitions do not meet these criteria, Soldiers should notify a supervisor.

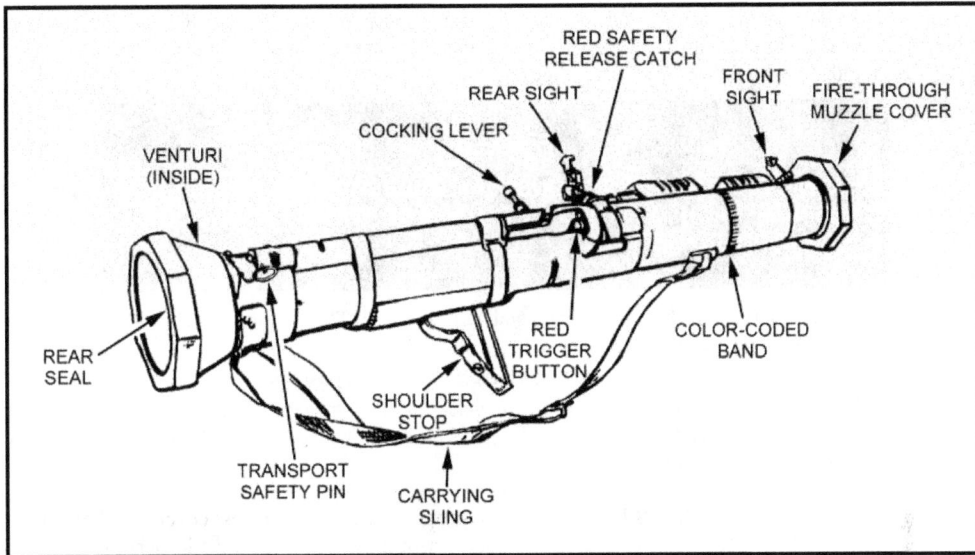

Figure 3-4. M136 AT4 inspection points.

M136A1 AT4 CONFINED SPACE SHOULDER-LAUNCHED MUNITION

3-16. M136A1 AT4CSs will usually be delivered to the supply point on pallets.

PACKAGING

3-17. Two M136A1 AT4CS munitions, each wrapped in a plastic barrier bag, are packed together in a wooden container. The containers are too heavy to stack more than four deep and two containers wide on the pallets (Figure 3-5).

3-18. Upon removal from the pallet, the wooden containers should be individually inspected for damage. During this inspection, the containers should meet the following criteria:

- The container is not punctured or damaged.
- The container has the correct markings, including the serial number and bar code on the end of the container.
- Fasteners are intact and show no signs of tampering.

3-19. If any of the containers do not meet these criteria, Soldiers should notify a supervisor.

MUNITION

3-20. Once a container passes inspection, it should be opened and the munition unpacked. To unpack the M136A1 AT4CS—

(1) Open the wooden container.
(2) Remove the plastic bags from the wooden container.
(3) Break the seal on the plastic bag.

Note. The plastic barrier bags have a V-notch cut 1 inch from the side to allow easy opening without the aid of a tool.

---

(4) Remove the munition.

3-21. The Soldier should then be able to observe and inspect the M136A1 AT4CS. The inspection is limited to a visual inspection of the outer tube and its components. During this inspection, the M136A1 AT4CS should meet the following criteria (Figure 3-6):

DANGER

USING DAMAGED MUNITIONS MAY CAUSE INJURY OR DEATH. INSPECT EACH MUNITION FOR DAMAGE WHEN RECEIVED. IF DAMAGED, DO NOT USE.

- The transport safety fork is in place and fully inserted. The lanyard is attached to the transport safety fork and the launcher.
- The cocking lever is present and in the SAFE (uncocked) position.
- The rear bumper and dust cover are in place and undamaged. Check the rear dust cover for moisture leaks (countermass leaks). If there is moisture/leakage coming from the rear dust cover, the munition cannot be fired. -23

Note. If the M136A1 AT4CS launcher has a damaged/punctured rear dust cover but shows no moisture/leakage, report it to a supervisor. A supervisor must then inspect the launcher to ensure that there is no damage to the countermass container and remove any debris.

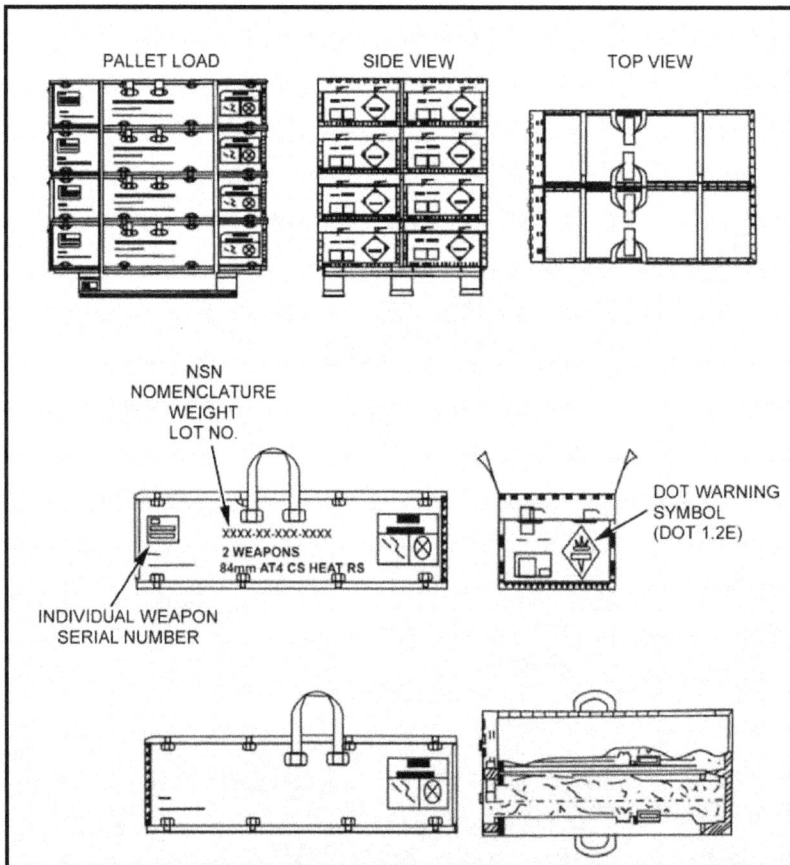

Figure 3-5. M136A1 AT4 confined space packaging and pallet details.

● The front bumper and dust cover are intact and no foreign objects are present. If the dust cover is torn, cut it out and check the launch tube to ensure it is clear of foreign objects. Remove any that you find by turning the tube muzzle downward and gently shaking the launcher.

Note. M136A1 AT4CS launchers with missing muzzle dust covers and no obstructions are suitable for use. However, these launchers should have their muzzle covers replaced as soon as possible to prevent further damage and deterioration.

● The launcher has the correct color-coded band.
● The sights function properly. Open the sight covers to ensure the sights pop up and are undamaged.
● The forward safety does not move when you depress it.
● The red trigger button is not missing.
● The launcher body has no cracks, dents, or bulges.
● The carrying sling is not frayed and is attached firmly to the launch tube.
● The front grip is attached firmly to the launcher. It remains closed until opened, and locks in the open position.
● The shoulder stop is not broken or damaged, and it unsnaps and folds down.

3-22. If any of the munitions do not meet these criteria, Soldiers should notify a supervisor.

Note. If the M136A1 AT4CS is not to be used immediately, it should be returned to its plastic bag and the bag resealed with tape.

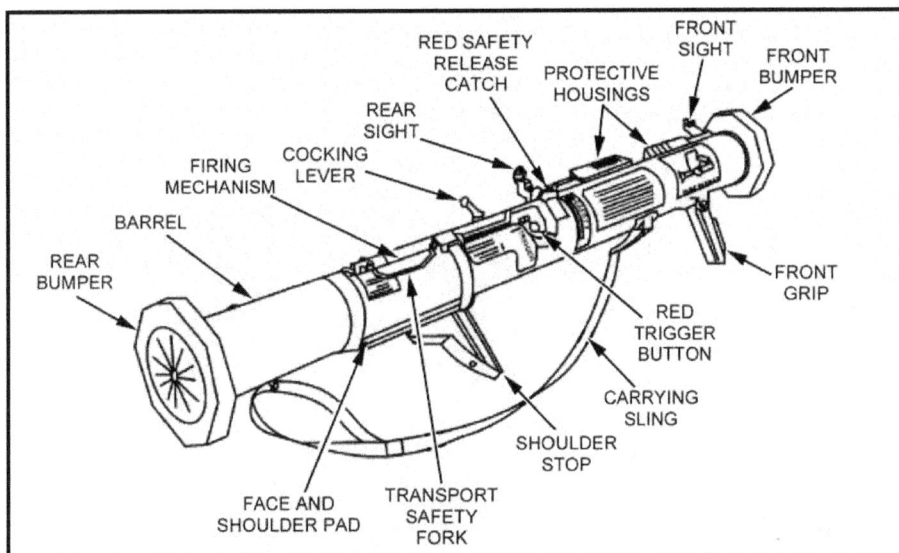

Figure 3-6. M136A1 AT4 confined space inspection points.

IMPROVED M72 SHOULDER-LAUNCHED MUNITION

3-23. Improved M72 shoulder-launched munitions will usually be delivered to the supply point on a pallet with a steel frame. They come encased in sealed individual MK14 aluminum containers (Figure 3-7).

3-24. To remove the improved M72 containers from the pallet—
    (1) Cut and remove the straps.
    (2) Remove the frame.
    (3) Remove three rows of containers.

(4) Cut and remove the upper girthwise strap.
(5) Remove two rows of containers.
(6) Cut and remove the lower girthwise strap.
(7) Remove the remaining containers.

PACKAGING

3-25. Upon removal from the pallet, the metal containers should be individually inspected for damage. During this inspection, the containers should meet the following criteria:

- The container is not punctured or damaged.
- The container has the correct markings, including the serial number and bar code on the side of the container.
- The lids are intact and show no signs of tampering.

3-26. If any of the containers do not meet these criteria, Soldiers should notify a supervisor.

MUNITION

3-27. Once a container passes inspection, it should be opened and the munition unpacked (Figure 3-8). To unpack the improved M72—

(1) Grasp the container, cut the wire seal, and unscrew the lid counterclockwise.
(2) Remove the cushion material and launcher.
(3) Save the container and cushion material for repacking the munition.

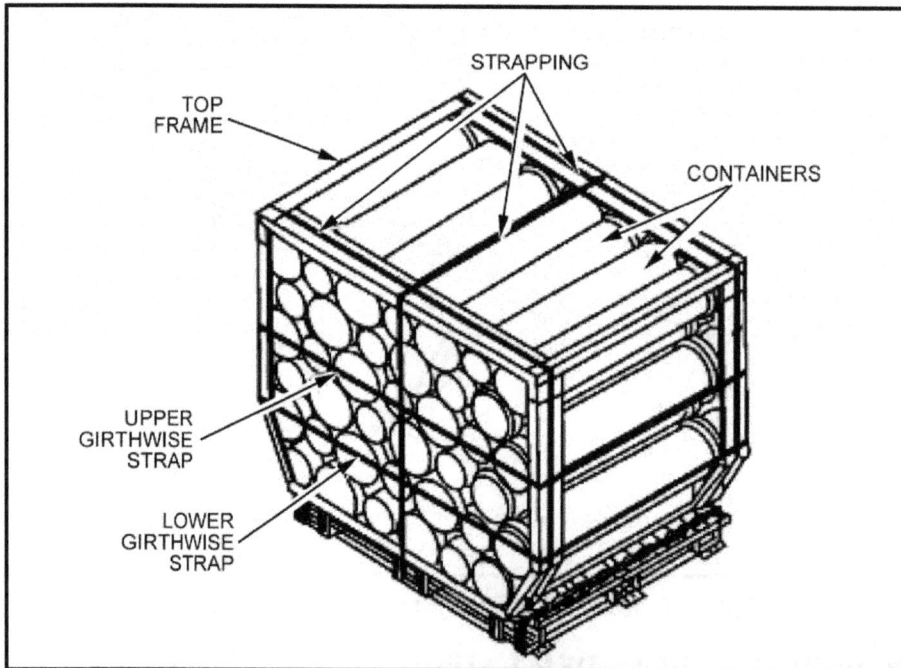

Figure 3-7. Improved M72 inspection and unpacking points.

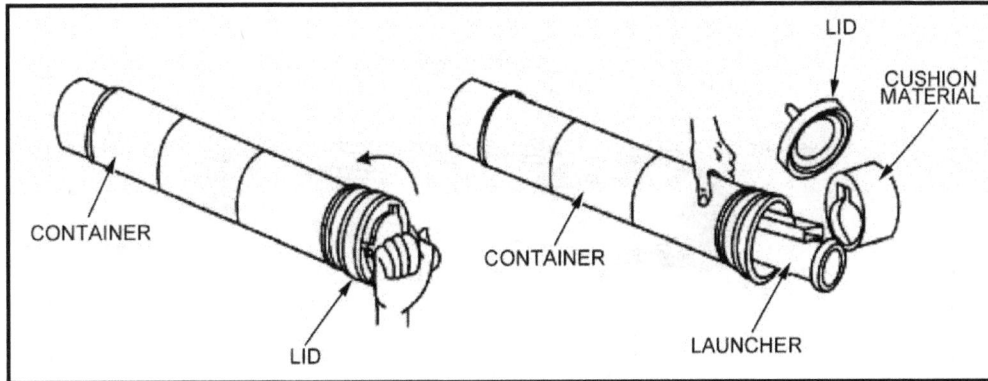

Figure 3-8. Improved M72 containers.

3-28. The Soldier should then be able to observe and inspect the munition. The inspection is limited to a visual inspection of the outer tube and its components. During this inspection, the munition should meet the following criteria (Figure 3-9):

> **DANGER**
>
> USING DAMAGED MUNITIONS MAY CAUSE INJURY OR DEATH. INSPECT EACH MUNITION FOR DAMAGE WHEN RECEIVED. IF DAMAGED, DO NOT USE.

- The launcher is not damaged, bent, or missing parts.
- The launch tube has no holes, gouges, cracks, bulges, or other damage.
- The sling is intact, and the sling hooks are not bent.
- The data plates and stickers are not blurred or missing, and do not contain incorrect nomenclature.
- The transport safety pin is in place and fully inserted.
- The lanyard is attached to the transport safety pin and the launcher.
- The transport pin lock is not broken.
- The trigger safety handle is present and in the SAFE (uncocked) position.
- The trigger cover is not torn.
- The igniter or flash tube is not split or separated from the launcher.
- The front cover and rear cover/shoulder stop is not damaged.
- The sights function properly. Extend the launcher to ensure the sights pop up and are undamaged.
- The launcher has the correct color-coded band.
- The launcher has no foreign debris present. Ensure the launcher is free of foreign debris before reattaching the sling.

Note. If the munition is not to be used immediately, it should be returned to its shipping container.

3-29. If any of the munitions do not meet these criteria, Soldiers should notify a supervisor.

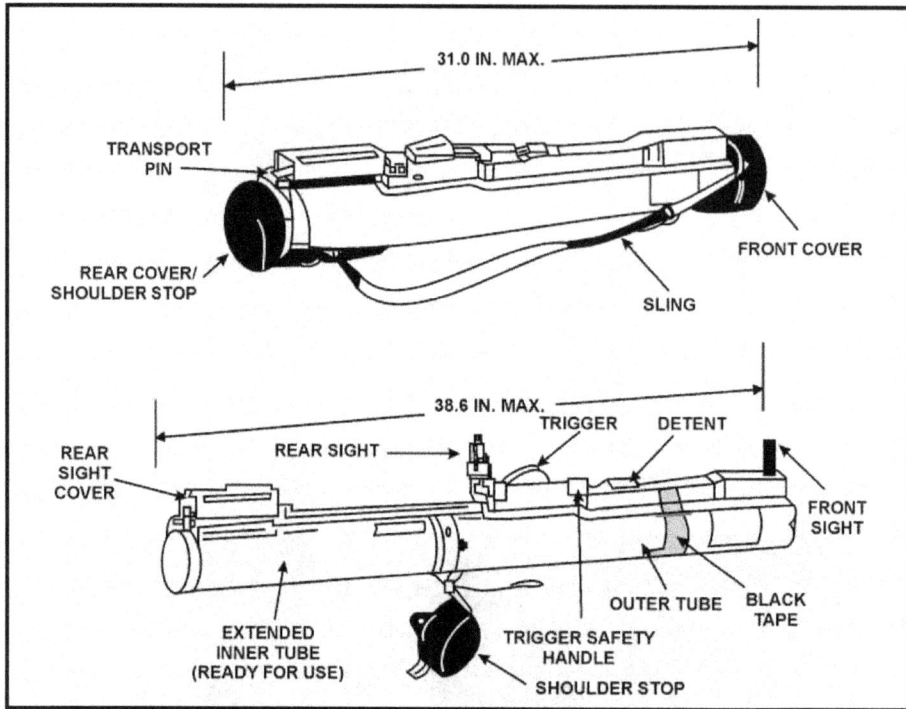

Figure 3-9. Improved M72 inspection points.

M72A2/3 SHOULDER-LAUNCHED MUNITION

3-30. M72A2/A3s will usually be delivered to the supply point in a wire-bound wooden box. Three inner packs are placed in the box (Figure 3-10).

3-31. To remove the M72A2/A3 launchers from the box—

    (1) Keep the marking "NOSE END" facing the least hazardous area.

    (2) Check for proper item nomenclature.

Note. Stand to the side of the "NOSE END" marking when opening the crate.

---

    (3) Unfasten the wire hoops.

WARNING

Do not use sharp instruments to open the crate cover or the inner package.

PACKAGING

3-32. Five complete M72A2/A3 munitions are packaged inside a box, within a fiberboard inner pack (Figure 3-11). The wooden boxes should be individually inspected for damage. During this inspection, the containers should meet the following criteria:

- The container is not punctured or damaged.

- The container has the correct markings, including the serial number and bar code on the end of the container.
- Fasteners are intact and show no signs of tampering.

3-33. If any of the containers do not meet these criteria, Soldiers should notify a supervisor.

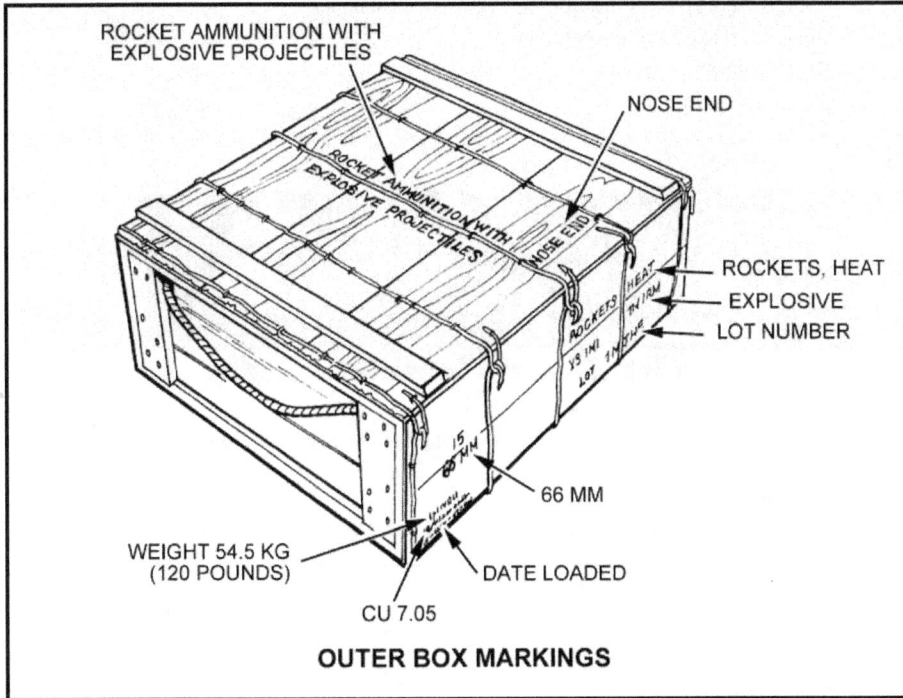

Figure 3-10. M72A2/A3 inspection and unpacking points.

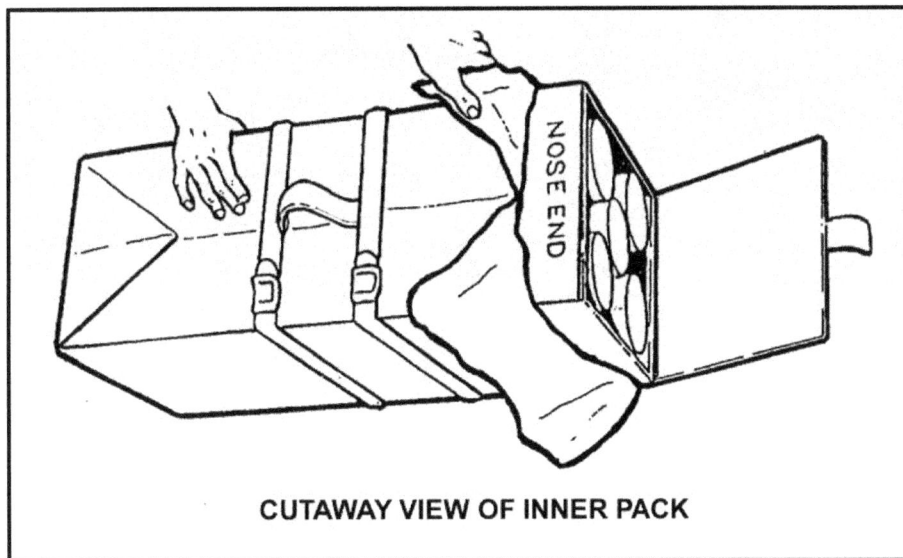

Figure 3-11. M72A2/A3 packaging details.

MUNITION

3-34. Once a container passes inspection, it should be opened and the munition unpacked. To unpack the M72A2/A3—

    (1) Open the wooden container.

    (2) Remove the fiberboard insert from the wooden container.

    (3) Break the seal on the fiberboard insert.

    (4) Remove the munition.

3-35. The Soldier should then be able to observe and inspect the munition. The inspection is limited to a visual inspection of the outer tube and its components. During this inspection, the munition should meet the following criteria (Figure 3-12):

> **DANGER**
>
> **USING DAMAGED MUNITIONS MAY CAUSE INJURY OR DEATH. INSPECT EACH MUNITION FOR DAMAGE WHEN RECEIVED. IF DAMAGED, DO NOT USE.**

- The body has no dents, cracks, or bulges.
- The rubber boots covering the trigger bar and barrel detent have no tears or punctures.
- The arming handle is present and in the SAFE position.
- The pull pin is in place.
- The data plate has the phrase "W/COUPLER (Figure 3-13)".

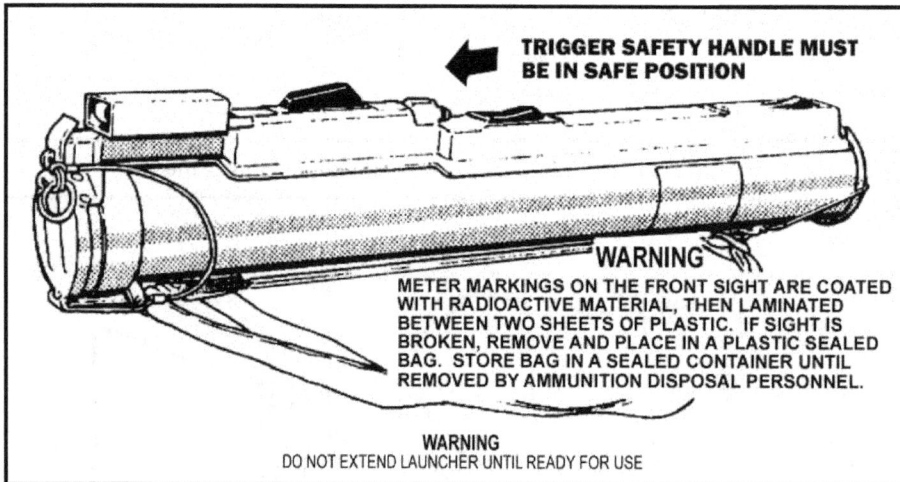

TRIGGER SAFETY HANDLE MUST BE IN SAFE POSITION

WARNING
METER MARKINGS ON THE FRONT SIGHT ARE COATED WITH RADIOACTIVE MATERIAL, THEN LAMINATED BETWEEN TWO SHEETS OF PLASTIC. IF SIGHT IS BROKEN, REMOVE AND PLACE IN A PLASTIC SEALED BAG. STORE BAG IN A SEALED CONTAINER UNTIL REMOVED BY AMMUNITION DISPOSAL PERSONNEL.

WARNING
DO NOT EXTEND LAUNCHER UNTIL READY FOR USE

Figure 3-12. M72A2/A3 inspection points.

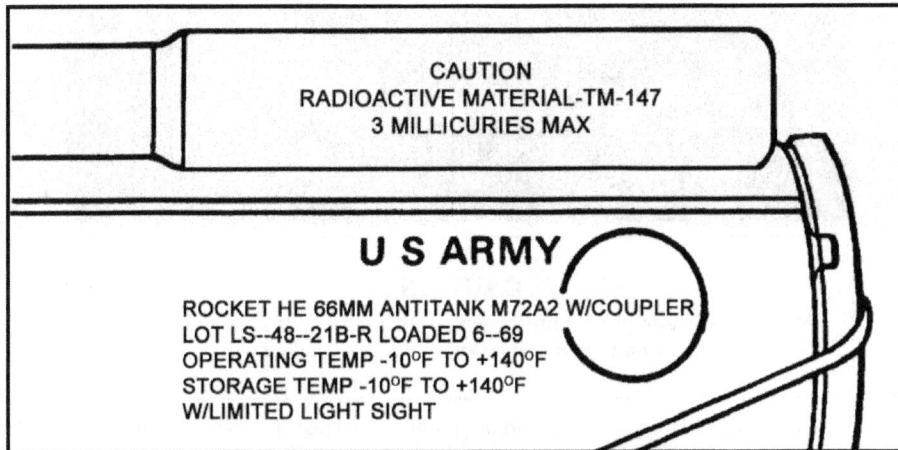

CAUTION
RADIOACTIVE MATERIAL-TM-147
3 MILLICURIES MAX

U S ARMY

ROCKET HE 66MM ANTITANK M72A2 W/COUPLER
LOT LS--48--21B-R LOADED 6--69
OPERATING TEMP -10°F TO +140°F
STORAGE TEMP -10°F TO +140°F
W/LIMITED LIGHT SIGHT

Figure 3-13. M72A2/A3 launcher data plate.

DANGER

IF THE M72A2 LAW DOES NOT STATE "W/COUPLER" ON ITS DATA PLATE, TURN THE MUNITION IN TO THE UNIT AMMUNITION SECTION. THE COUPLER PREVENTS THE INNER AND OUTER TUBES FROM SEPARATING AND POSSIBLY CAUSING PREMATURE DETONATION.

Note. If the munition is not to be used immediately, it should be returned to its shipping container.

3-36. If any of the munitions do not meet these criteria, Soldiers should notify a supervisor.

SECTION II. ARMING PROCEDURES

All shoulder-launched munitions require some form of preparation before firing. The following paragraphs outline the steps involved in preparing each type.

M141 BUNKER DEFEAT MUNITION

3-37. To prepare the M141 BDM—

WARNING

Keep the munition pointed toward the target, and keep the backblast area clear.

DANGER

WHEN OPERATING A SHOULDER-LAUNCHED MUNITION, KEEP IT POINTED IN THE DIRECTION OF THE TARGET. ENSURE YOUR WHOLE BODY IS CLEAR OF THE MUZZLE AND REAR OF THE LAUNCHER, AND ENSURE THE BACKBLAST AREA IS CLEAR.

CAUTION

Always wear ear protection, ballistic eye protection, a helmet, and a protective vest when firing shoulder-launched munitions.

(1) Remove the M141 BDM from its carrying position, and cradle it in your left arm (Figure 3-14).

Figure 3-14. Cradling the M141 bunker defeat munition.

(2) Keeping the munition's muzzle pointed toward the target area, perform the following step:
- In *training*, face the rear when preparing the round for firing.
- In *combat*, keep your eyes on the target area while preparing the round for firing. Ensure that nearby Soldiers are aware of your intent, and that the backblast area is clear before firing.

(3) Supporting the munition with your right arm, place the munition under your left arm.

(4) With your right hand, pull and release the transport safety pin (Figure 3-15).

Note. This pin is important; you must reinsert it if you do not fire the launcher. Unless it is attached to the launcher with a lanyard, you must keep it in a safe place.

(5) Depress the tube release button with your left thumb (Figure 3-16).

(6) Grasp the rear tube (inner tube) just in front of the rear bumper with your right hand, and extend the inner tube rearward until it stops.

Note. A yellow band is visible at the inner tube front end when the tube is fully extended.

(7) Release the tube release button.

(8) Rotate the inner tube clockwise (in the direction of the arrow) until it locks (Figure 3-17).

Figure 3-15. Removing the transport safety pin on the M141 bunker defeat munition.

Figure 3-16. Extending the tube on the M141 bunker defeat munition.

Figure 3-17. Locking the M141 bunker defeat munition inner tube.

(9) Verify that the inner tube is locked by attempting to rotate the inner tube counterclockwise (opposite to the arrow).

Note. If the tubes are not locked, the munition will not arm.

(10) Inspect the inner tube for cracks, dents, or punctures. If any are present, return the munition to the carrying configuration, and dispose of it according to the unit SOP.

(11) Press the shoulder stop lock/release button, and pull the shoulder stop out (Figure 3-18).

Figure 3-18. Unlocking and unfolding the shoulder stop on the M141 bunker defeat munition.

(12) Grip the forward end of the munition with your left hand and the rear end of the munition with your right hand.

(13) Raise the munition out and away from your body.

(14) While keeping the munition pointed at the target, pivot your body 180 degrees to face the target.

(15) Place the munition on your right shoulder.

(16) Reach forward with your left hand, and grasp the front sight cover. Press down, and slide it rearward (Figure 3-19).

(17) With your left hand, grasp the rear sight cover. Press down, and slide it forward (Figure 3-19).

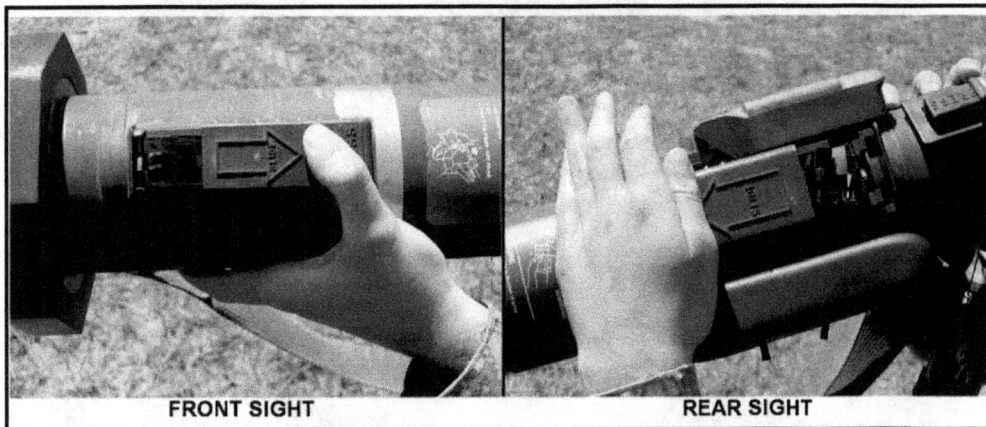

FRONT SIGHT   **REAR SIGHT**

Figure 3-19. Deploying the sights on the M141 bunker defeat munition.

(18) Wrap the sling strap around your left bicep. Cup the bottom of the munition with your left hand, and slide it back toward your body to tighten the sling.

Note. When firing the M141 BDM, the weapon sling should be used to increase firer control, as is done with a conventional rifle; however, DO NOT wrap the sling around your left arm as one would with a rifle.

WARNING

Check the backblast area before firing the munition.

---

(19) Ensure the backblast area is clear of personnel.

(20) Grasp the firing mechanism cover with your right hand, and rotate the cover all the way forward until the cover is flush with the outer tube (Figure 3-20).

Notes. 1. If the firing mechanism cover is not flush with the launch tube, the munition will not arm.

2. The word ARMED can be seen in red letters when the cover is opened.

(21) Adjust the rear sight to the correct range, using the following (Figure 3-21):

Note. When opening the rear sight cover, the range is preset at the 150-meter battlesight range setting.

- To adjust the rear sight range setting to more than 150 meters, turn the range knob clockwise (toward the muzzle).
- To decrease the range, turn the range knob counterclockwise (toward the firer).

Note. There is an audible clicking sound at each 50-meter increment; this sound aids you during limited visibility.

Figure 3-20. Arming the firing mechanism on the M141 bunker defeat munition.

Figure 3-21. Adjusting the rear sight on the M141 bunker defeat munition.

(22) Place the fingertips of your right hand on the safety button (located on top of the firing mechanism), and press down. Then, place your right thumb on the red trigger button.

(23) Pull the shoulder stop against your shoulder.

(24) Aim the launcher.

Note. The rear sight should be no less than 2 1/2 inches and no more than 3 inches from your eyes.

(25) Press the trigger button forward with the thumb of your right hand, and hold until the munition fires (Figure 3-22).

Figure 3-22. Firing the M141 bunker defeat munition.

M136 AT4 SHOULDER-LAUNCHED MUNITION

3-38. To prepare the M136 AT4—

WARNING

Keep the munition pointed toward the target, and keep the backblast area clear.

DANGER

WHEN OPERATING A SHOULDER-LAUNCHED MUNITION, KEEP IT POINTED IN THE DIRECTION OF THE TARGET. ENSURE YOUR WHOLE BODY IS CLEAR OF THE MUZZLE AND REAR OF THE LAUNCHER, AND ENSURE THE BACKBLAST AREA IS CLEAR.

CAUTION

Always wear ear protection, ballistic eye protection, a helmet, and a protective vest when firing shoulder-launched munitions.

---

(1) Remove the M136 AT4 from its carrying position, and cradle it in your left arm (Figure 3-23).

(2) Keeping the munition's muzzle pointed toward the target area, perform the following step:

- In *training*, face to your right when preparing the round for firing.
- In *combat*, keep your eyes on the target area while preparing the round for firing. Ensure that nearby Soldiers are aware of your intent, and that the backblast area is clear before firing.

(3) With your right hand, pull and release the transport safety pin (Figure 3-24).

Note. This pin is important; you must reinsert it if you do not fire the launcher. Unless it is attached to the launcher with a lanyard, you must keep it in a safe place.

(4) Unsnap, unfold, and hold the shoulder stop with your right hand (Figure 3-25).
(5) Grip the base of the sling on the front of the launcher with your left hand and the shoulder stop with your right hand.
(6) Raise the munition out and away from your body.
(7) While keeping the munition pointed at the target, pivot your body 90 degrees to face the target.
(8) Place the munition on your right shoulder.

Note. You can use the carrying strap to steady the munition.

Figure 3-23. Cradling the M136 AT4.

Figure 3-24. Removing the transport safety pin on the M136 AT4.

Figure 3-25. Unsnapping the shoulder stop on the M136 AT4.

(9) Reach forward with your right hand, and grasp the front sight cover. Press down, and slide it rearward (Figure 3-26).

(10) With your right hand, grasp the rear sight cover. Press down, and slide it forward (Figure 3-26).

WARNING

Check the backblast area before you cock the launcher.

---

(11) Ensure the backblast area is clear of personnel.

(12) Unfold the cocking lever with your right hand (Figure 3-27). Place your thumb under it and, with the support of your fingers in front of the firing mechanism, push it forward, rotate it downward and to the right, and let it slide backward.

Figure 3-26. Opening the sights on the M136 AT4.

Figure 3-27. Cocking the M136 AT4.

(13) Adjust the rear sight to the correct range, using the following (Figure 3-28):

Note. When opening the rear sight cover, the range is preset at the 200-meter battlesight range setting.

- To adjust the rear sight range setting to more than 200 meters, turn the range knob clockwise (toward the muzzle).
- To decrease the range, turn the range knob counterclockwise (toward the firer).

Note. There is an audible clicking sound at each 50-meter increment; this sound aids you during limited visibility.

Figure 3-28. Adjusting the rear sight on the M136 AT4.

(14) Place the first two fingers of your right hand on the red safety release catch, and extend the thumb (Figure 3-29). While keeping the thumb extended, press the red safety release catch down, and hold.

(15) Pull back on the sling with your left hand to seat the shoulder stop firmly against your shoulder.

(16) Aim the launcher.

Note. The rear sight should be no less than 2 1/2 inches and no more than 3 inches from your eyes.

(17) Press the red trigger button with the thumb of your right hand to fire the launcher, and hold until the munition fires (Figure 3-30).

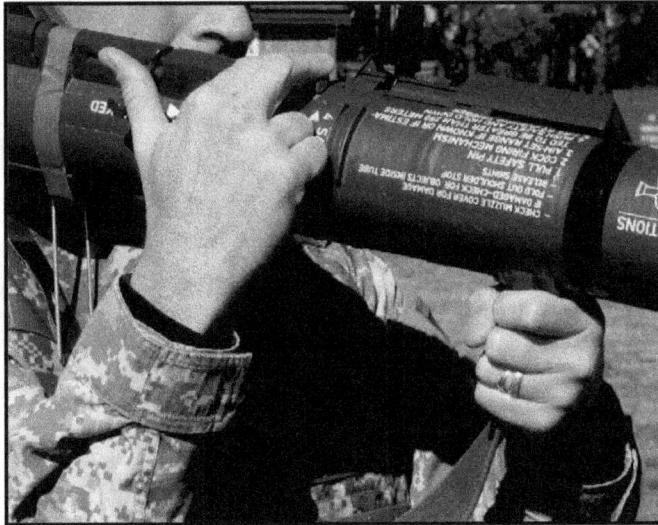

Figure 3-29. Releasing the red safety release catch on the M136 AT4.

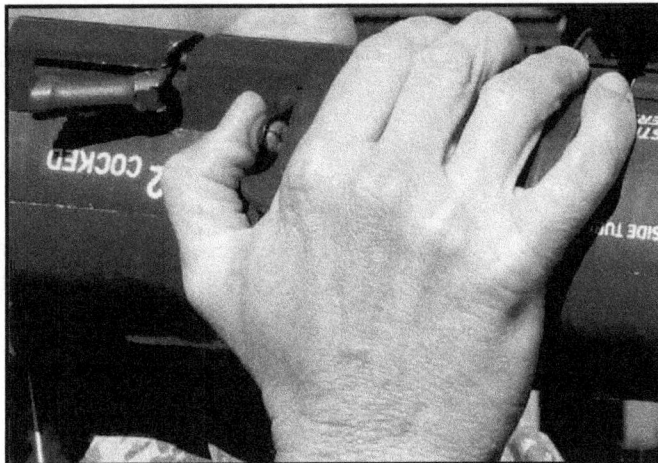

Figure 3-30. Pressing the red trigger button to fire the M136 AT4.

M136A1 AT4 CONFINED SPACE SHOULDER- LAUNCHED MUNITION

3-39. To prepare the M136A1 AT4CS—

WARNING

Keep the munition pointed toward the target, and keep the backblast area clear.

DANGER

WHEN OPERATING A SHOULDER-LAUNCHED MUNITION, KEEP IT POINTED IN THE DIRECTION OF THE TARGET. ENSURE YOUR WHOLE BODY IS CLEAR OF THE MUZZLE AND REAR OF THE LAUNCHER, AND ENSURE THE BACKBLAST AREA IS CLEAR.

CAUTION

Always wear ear protection, a helmet, ballistic eye protection, and a protective vest when firing shoulder-launched munitions.

---

(1) Remove the M136A1 AT4CS from its carrying position, and cradle it in your left arm (Figure 3-31).

Figure 3-31. Cradling the M136A1 AT4 confined space.

(2) Keeping the munition's muzzle pointed toward the target area, perform the following step:

- In *training*, face to your right when preparing the round for firing.
- In *combat*, keep your eyes on the target area while preparing the round for firing. Ensure that nearby Soldiers are aware of your intent and that the backblast area is clear before firing.

(3) With your right hand, pull and release the transport safety fork (Figure 3-32).

Note. This fork is important; you must reinsert it if you do not fire the launcher. Unless it is attached to the launcher with a lanyard, you must keep it in a safe place.

Figure 3-32. Removing the transport safety fork on the M136A1 AT4 confined space.

(4) Unsnap and unfold the shoulder stop with your right hand (Figure 3-33).

Figure 3-33. Unsnapping the shoulder stop on the M136A1 AT4 confined space.

(5) Unfold the front grip with your right hand (Figure 3-34).

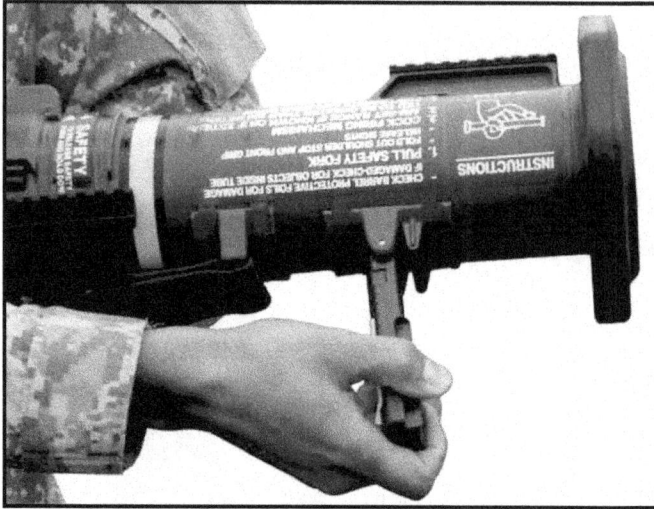

Figure 3-34. Unfolding the front grip on the M136A1 AT4 confined space.

(6) Grip the shoulder stop with your right hand and the front grip with your left hand.

(7) Raise the munition out and away from your body.

(8) While keeping the munition pointed at the target, pivot your body 90 degrees to face the target.

(9) Place the munition on your right shoulder.

Note. You can use the front grip (Figure 3-35) or carrying strap to steady the munition.

USING THE FRONT GRIP   USING THE CARRYING STRAP

Figure 3-35. Stabilizing the M136A1 AT4 confined space.

(10) While supporting the munition with your left hand by holding the front grip/forward sling, reach forward with your right hand, and grasp the front sight cover. Press down, and slide it rearward (Figure 3-36).

Figure 3-36. Opening the front sight cover on the M136A1 AT4 confined space.

(11) With your right hand, grasp the rear sight cover. Press down, and slide it forward (Figure 3-37).

Figure 3-37. Opening and adjusting the sights on the M136A1 AT4 confined space.

WARNING

Check the backblast area before you cock the launcher.

---

(12) Ensure the backblast area is clear of personnel.

(13) Unfold the cocking lever with your right hand (Figure 3-38). Place your thumb under it and, with the support of your fingers in front of the firing mechanism, push it forward, rotate it downward and to the right, and let it slide backward.

Figure 3-38. Cocking the M136A1 AT4 confined space.

(14) Adjust the rear sight to the correct range, using the following (Figure 3-39):

Note. When opening the rear sight cover, the range is preset at the 200-meter battlesight range setting.

- To adjust the rear sight range setting to more than 200 meters, turn the range knob clockwise (toward the muzzle).
- To decrease the range, turn the range knob counterclockwise (toward the firer).

Note. There is an audible clicking sound at each 50-meter increment; this sound aids you during limited visibility.

(15) Place the first two fingers of your right hand on the red safety release catch, and extend your thumb (Figure 3-40). While keeping your thumb extended, press the red safety release catch down, and hold.

(16) Pull back on the front grip/sling with your left hand to seat the shoulder stop firmly against your shoulder, and hold.

(17) Aim the launcher.

Note. The rear sight should be no less than 2 1/2 inches and no more than 3 inches from your eyes.

(18) Press the red trigger button with the thumb of your right hand to fire the launcher, and hold until the munition fires (Figure 3-41).

Figure 3-39. Adjusting the rear sight on the M136A1 AT4 confined space.

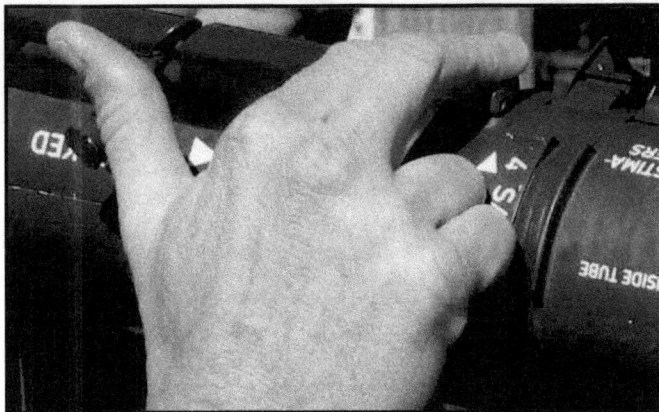

Figure 3-40. Releasing the red safety catch on the M136A1 AT4 confined space.

Figure 3-41. Pressing the red trigger button to fire the M136A1 AT4 confined space.

IMPROVED M72 SHOULDER-LAUNCHED MUNITION

3-40. To prepare the improved M72—

Note. The improved M72 shoulder-launched munition can be fired from your left or right shoulder.

WARNING

Keep the munition pointed toward the target, and keep the backblast area clear.

DANGER

WHEN OPERATING A SHOULDER-LAUNCHED MUNITION, KEEP IT POINTED IN THE DIRECTION OF THE TARGET. ENSURE YOUR WHOLE BODY IS CLEAR OF THE MUZZLE AND REAR OF THE LAUNCHER, AND ENSURE THE BACKBLAST AREA IS CLEAR.

CAUTION

Always wear ear protection, ballistic eye protection, a helmet, and a protective vest when firing shoulder-launched munitions.

---

(1) Remove the improved M72 from its carrying position, and cradle it in your left arm (Figure 3-42).

Figure 3-42. Cradling the improved M72.

(2) Remove the transport safety pin (Figure 3-43).

Figure 3-43. Removing the improved M72 transport safety pin.

(3) Rotate the rear cover downward so the front cover and adjustable sling assembly can fall free (Figures 3-44). Pull the sling away with the nonfiring hand, if needed.

Note. Do not discard the sling assembly until after you fire the rocket.

Figure 3-44. Releasing the improved M72 rear cover/shoulder stop.

(4) With your firing hand, grasp the rear sight cover. With your nonfiring hand, grasp the launcher forward of the barrel detent. Pull your hands sharply in opposite directions to extend the launcher (Figure 3-45).

Figure 3-45. Extending the improved M72 launcher.

(5) Ensure that the launcher is fully extended and locked by trying to close it.

(6) Grip underneath the rear end of the launcher with your right hand and underneath the forward end of the launcher with your left hand.

(7) Raise the munition out and away from your body.

(8) While keeping the munition pointed at the target, pivot your body 180 degrees to face the target.

(9) Place the munition on your right shoulder.

WARNING

Check the backblast area before you arm the launcher.

---

(10) Ensure the backblast area is clear of personnel.

(11) Pull the trigger arming handle to the ARM position.

Note. If the trigger arming handle will not remain in the ARM position, the launcher is not fully extended.

(12) Adjust the rear sight to the correct range, using the following (Figure 3-46):

Note. When opening the rear sight cover, the range is preset at the lowest range setting of 50 meters.

- To adjust the rear sight range setting to more than 50 meters, turn the range knob clockwise (toward the muzzle).
- To decrease the range, turn the range knob counterclockwise (toward the firer).

Note. There is an audible clicking sound at each 50-meter increment; this sound aids you during limited visibility.

Figure 3-46. Adjusting the rear sight on the improved M72.

(13) Pull the rear cover/shoulder stop firmly against your shoulder.

(14) Aim the launcher.

Note. The rear sight should be no less than 2 1/2 inches and no more than 3 inches from your eyes.

(15) Depress the rubber boot on the trigger bar firmly with the fingers of your firing hand, and hold until the munition fires (Figure 3-47).

Figure 3-47. Firing the improved M72.

M72A2/A3 SHOULDER-LAUNCHED MUNITION

3-41. To prepare the M72A2/A3—

Note. The M72A2/A3 shoulder-launched munition can be fired from your left or right shoulder.

WARNING

Keep the munition pointed toward the target, and keep the backblast area clear.

DANGER

WHEN OPERATING A SHOULDER-LAUNCHED MUNITION, KEEP IT POINTED IN THE DIRECTION OF THE TARGET. ENSURE YOUR WHOLE BODY IS CLEAR OF THE MUZZLE AND REAR OF THE LAUNCHER, AND ENSURE THE BACKBLAST AREA IS CLEAR.

CAUTION

Always wear ear protection, ballistic eye protection, a helmet, and a protective vest when firing shoulder-launched munitions.

---

(1) Remove the M72A2/A3 from its carrying position, and cradle it in your left arm (Figure 3-48).

Figure 3-48. Cradling the M72A2/A3.

(2) Remove the transport safety pin (Figure 3-49).

Figure 3-49. Removing the M72A2/A3 transport safety pin.

(3) Rotate the rear cover downward so the front cover and adjustable sling assembly can fall free (Figure 3-50). Pull the sling away with the nonfiring hand, if needed.

Note. Do not discard the sling assembly until after you fire the rocket.

Figure 3-50. Removing the M72A2/A3 front cover and the adjustable sling assembly.

(4) With your firing hand, grasp the rear sight cover. With your nonfiring hand, grasp the launcher forward of the barrel detent. Pull your hands sharply in opposite directions to extend the launcher (Figure 3-51).

Figure 3-51. Extending the M72A2/A3 launcher.

(5) Ensure that the launcher is fully extended and locked by trying to close it.

(6) Grip underneath the rear end of the launcher with your right hand and underneath the forward end of the launcher with your left hand.

(7) Raise the munition out and away from your body.

(8) While keeping the munition pointed at the target, pivot your body 90 degrees to face the target.

(9) Place the munition on your right shoulder.

Note. M72-series munitions can be fired from your left or right shoulder.

WARNING

Check the backblast area before you cock the launcher.

---

(10) Ensure the backblast area is clear of personnel.

(11) Pull the trigger arming handle to the ARM position.

Note. If the trigger arming handle will not remain in the ARM position, the launcher is not fully extended.

(12) Pull the rear cover/shoulder stop firmly against your shoulder, and hold.

(13) Aim the launcher.

Note. Place the sight at an easy reading distance.

(14) Press the red trigger button with the thumb of your right hand, and hold until the munition fires (Figure 3-52).

Figure 3-52. Firing the M72A2/A3.

SECTION III. OPERATION UNDER UNUSUAL CONDITIONS

3-42. Soldiers must be able to operate shoulder-launched munitions under various operational conditions. Further, Soldiers may also encounter events that change normal operation of the munition, such as misfires.

MISFIRE PROCEDURES

3-43. A misfire is a complete failure to fire caused by a procedural or mechanical failure. A misfire is not dangerous, but since it cannot be immediately distinguished from a delay in the functioning of the firing mechanism, it should be considered as a possible hangfire until such possibility has been eliminated. The procedures used to address a misfire may vary according to the Soldier's environment—combat or training.

DANGER

WHEN OPERATING A SHOULDER-LAUNCHED MUNITION, KEEP IT POINTED IN THE DIRECTION OF THE TARGET. ENSURE YOUR WHOLE BODY IS CLEAR OF THE MUZZLE AND REAR OF THE LAUNCHER, AND ENSURE THE BACKBLAST AREA IS CLEAR. NOTIFY THE EOD UNIT WHEN THERE IS A MISFIRE. DO NOT RETURN MISFIRED LAUNCHERS TO THE ASP.

Note. Notify your supervisor and ASP of any unusual occurrence, regardless of whether the munition fires or not. Examples include excessive overpressure, recoil, or heat on your face after you have fired the munition.

M141 BUNKER DEFEAT MUNITION

3-44. On the M141 BDM, a misfire is usually caused by one of the following factors:
- The firing mechanism is not armed.
- The inner tube is not fully extended and locked.
- The firing mechanism or the propelling charge explosive train is faulty.

Combat Environment

3-45. To address a misfire that occurs in combat—

WARNING

Keep the munition pointed toward the target, and keep the backblast area clear.

---

(1) Release the trigger button and safety button.
(2) Squeeze the safety button firmly, and hold, aim, and press the trigger button a second time.
(3) If the munition does not fire, announce "MISFIRE" just loud enough for friendly personnel in the immediate area to hear.
(4) Release the trigger and safety buttons, and close the firing mechanism cover (SAFE position).
(5) Check the backblast area, and open the firing mechanism cover again (flush with the tube).
(6) Squeeze the safety button, and hold, aim, and press the trigger button a third time.
(7) If the munition still does not fire, say "MISFIRE" just loud enough for friendly personnel in the immediate area to hear.
(8) Release the trigger and safety buttons, and close the firing mechanism cover (SAFE position).
(9) If time permits, wait 90 seconds. Remove the munition from your shoulder, keeping the munition pointed toward the target, and cradle the munition in your left arm.
(10) Break off the sights to identify the misfired launcher.
(11) DO NOT collapse the launcher. Carefully lay your munition on the ground facing the target. Notify your supervisor.

Notes. 1. If the tactical situation permits, move to another location, and prepare another shoulder-launched munition.

2. As soon as you can, dispose of the misfired launcher in accordance with the unit SOP.

Training Environment

3-46. To address a misfire that occurs on the live-fire training range—

WARNING

Keep the munition pointed toward the target, and keep the backblast area clear.

---

(1) Release the trigger button and safety button.
(2) Squeeze the safety button firmly, and hold, aim, and press the trigger button a second time.
(3) If the munition still does not fire, announce "MISFIRE."
(4) Release the trigger and safety buttons, and close the firing mechanism cover (SAFE position).
(5) Check the backblast area, and open the firing mechanism cover again (flush with the tube).
(6) Squeeze the safety button, and hold, aim, and press the trigger button a third time.
(7) If the munition still does not fire, announce "MISFIRE."
(8) Release the trigger and safety buttons, and close the firing mechanism cover (SAFE position).
(9) Wait 90 seconds; remove the munition from your shoulder. DO NOT collapse the launcher. Carefully lay your munition on the ground facing the target. Notify range cadre.

Notes. 1. Wait two minutes before moving the munition from the firing line.

2. Notify post range control of the training situation, and follow local safety SOPs and regulations.

M136-SERIES SHOULDER-LAUNCHED MUNITIONS

3-47. On M136-series shoulder-launched munitions, misfires are usually caused by one of the following factors:

- The safety release catch is not depressed far enough to disengage the safety.
- The firing mechanism is faulty.
- The firing mechanism or the propelling charge explosive train is faulty.

Combat Environment

3-48. To address a misfire that occurs in combat—

WARNING

Keep the munition pointed toward the target, and keep the backblast area clear.

---

(1) Say "MISFIRE" just loud enough for friendly personnel in the immediate area to hear, while maintaining the original sight picture.
(2) Release the red trigger button and the red safety release catch.
(3) If time permits, wait five seconds. Remove your right hand from the firing mechanism, check the backblast area, and cock the munition again.
(4) Press down on the red safety release catch firmly, and hold. Aim the munition. Press and hold the red trigger button.
(5) If the munition does not fire, say "MISFIRE" just loud enough for friendly personnel in the immediate area to hear.
(6) Release the red trigger button and red safety release catch.
(7) If time permits, maintain the firing position for two minutes, and return the cocking lever to the SAFE (uncocked) position.
(8) Remove the munition from your shoulder, keeping the munition pointed toward the target.
(9) Cradle the munition in your left arm.
(10) Reinsert the transport safety pin/fork.
(11) Break off the sights to identify the misfired launcher.
(12) Carefully lay your munition on the ground facing the target. Notify your supervisor.

Notes. 1. If the tactical situation permits, move to another location, and prepare another shoulder-launched munition.

2. If the transport safety pin/fork cannot be reinserted or if the pin/fork is missing, notify the EOD unit.

3. As soon as you can, dispose of the misfired launcher in accordance with the unit SOP.

Training Environment

3-49. To address a misfire that occurs on the live-fire training range—

WARNING

Keep the munition pointed toward the target, and keep the backblast area clear.

---

(1) If the munition does not fire, announce "MISFIRE."

(2) Release the red trigger button, and the red safety release catch.

(3) Wait five seconds. Remove your right hand from the firing mechanism, check the backblast area, and cock the munition again.

Note. Count the seconds by saying "one thousand and one, one thousand and two" and so on.

---

(4) Press down on the red safety release catch firmly, and hold. Aim the munition. Press and hold the red trigger button.

(5) If the munition does not fire, say "MISFIRE."

(6) Release the red trigger button and red safety release catch.

(7) Maintain the firing position for two minutes, and return the cocking lever to the SAFE (uncocked) position.

(8) Remove the munition from your shoulder, keeping the munition pointed toward the target.

(9) Cradle the munition in your left arm.

(10) Reinsert the transport safety pin/fork.

(11) Carefully lay your munition on the ground facing the target. Notify range cadre.

Notes.   1.   If the transport safety pin/fork cannot be reinserted or if the pin is missing, notify the EOD unit.

         2.   Wait two minutes before moving the munition from the firing line.

         3.   Notify post range control of the training situation, and follow local safety SOPs and regulations.

M72-Series Shoulder-Launched Munitions

3-50. On M72-series shoulder-launched munitions, misfires are usually caused by one of the following factors:

- The firing mechanism is not armed.
- The inner tube is not fully extended and locked.
- The firing mechanism or the propelling charge explosive train is faulty.

Combat Environment

3-51. To address a misfire that occurs in combat—

WARNING

Keep the munition pointed toward the target, and keep the backblast area clear.

---

(1) Release the trigger, and squeeze the trigger spring boot again.

(2) If the munition still fails to fire, say "MISFIRE" just loud enough for friendly personnel in the immediate area to hear.

(3) Maintain your firing position for ten seconds, and place the trigger safety handle on SAFE.

Note. Count the seconds by saying "one thousand and one, one thousand and two" and so on.

---

(4) If time permits, wait one minute. Remove the munition from your shoulder, keeping it pointed in the direction of the target.

(5) Push the detent and grab the rear sight cover to partially collapse the launcher (about 4 inches). Extend it to the locked position, and push in on the launcher to ensure it is fully locked.

(6) Place the launcher on your shoulder; check the backblast area again; and then arm, aim, and fire the munition.

(7) If the launcher fails to fire, firmly squeeze the trigger spring boot again.

(8) If the munition still does not fire, say "MISFIRE" just loud enough for friendly personnel in the immediate area to hear.

(9) Release the trigger. Maintain your firing position for ten seconds, and place the trigger safety handle on SAFE.

(10) If time permits, wait one minute. Remove the munition from your shoulder, keeping the munition pointed toward the target, and cradle the munition in your left arm.

(11) Break off the sights to identify the misfired munition. DO NOT collapse the launcher.

(12) Carefully lay the munition on the ground facing the target.

(13) Notify your supervisor.

Notes. 1. If the tactical situation permits, move to another location, and prepare another shoulder-launched munition.

2. As soon as you can, dispose of the misfired launcher in accordance with the unit SOP.

Training Environment

3-52. To address a misfire that occurs on the live-fire training range—

WARNING

Keep the munition pointed toward the target, and keep the backblast area clear.

---

(1) Release the trigger, and squeeze the trigger spring boot again.

(2) If the munition still fails to fire, announce "MISFIRE."

(3) Maintain your firing position for ten seconds, and place the trigger safety handle on SAFE.

Note. Count the seconds by saying "one thousand and one, one thousand and two" and so on.

(4) Remove the munition from your shoulder, keeping it pointed in the direction of the target.

(5) Wait one minute. Partly collapse the launcher (about four inches), and then extend it to the locked position. Push in on the launcher to ensure it is fully locked.

(6) Place the launcher on your shoulder; check the backblast area again; and then arm, aim, and fire the munition.

(7) If the launcher fails to fire, firmly squeeze the trigger spring boot again.

(8) If the munition still does not fire, announce "MISFIRE."

(9) Release the trigger. Maintain your firing position for ten seconds, and place the trigger safety handle on SAFE.

(10) Wait one minute. Remove the munition from your shoulder, keeping the munition pointed toward the target. DO NOT collapse the launcher.

(11) Carefully lay the munition on the ground. Notify range cadre.

Notes.   1.   Wait two minutes before moving the munition from the firing line.

           2.   Notify post range control of the training situation, and follow local safety SOPs and regulations.

RESTORING TO A CARRYING CONFIGURATION

3-53. If the launcher is prepared to fire, but then is not fired, it should be returned to the carrying configuration to permit safe storage, transportation, and use at a later time. This ensures the physical integrity of the round and reduces the likelihood of damage from moisture and debris.

> **DANGER**
>
> WHEN OPERATING A SHOULDER-LAUNCHED MUNITION, KEEP IT POINTED IN THE DIRECTION OF THE TARGET. ENSURE YOUR WHOLE BODY IS CLEAR OF THE MUZZLE AND REAR OF THE LAUNCHER, AND ENSURE THE BACKBLAST AREA IS CLEAR.

Note. Once the launcher is returned to the carrying configuration, the Soldier can use the carrying strap for transport or cradle the munition in his arms for transport over short distances.

> **WARNING**
>
> Never use the sling to carry any shoulder-launched munition while in the extended configuration.

M141 BUNKER DEFEAT MUNITION

3-54. To restore the M141 BDM to a carrying configuration—

   (1) Release the safety button.
   (2) Close the firing mechanism cover.
   (3) Return the rear sight to the battlesight setting (150 meters).

WARNING

Do not close the rear sight cover before ensuring the rear sight is set to 150 meters. This can severely damage the sight.

---

   (4) Fold the rear sight down, hold it down, and close the rear sight cover.
   (5) Fold the front sight down, hold it down, and close the front sight cover.
   (6) Grip the forward end of the launcher with your left hand and the rear end of the launcher with your right hand.
   (7) Remove the munition from your shoulder, rotate your body 180 degrees (keeping the munition pointed downrange), and hold the munition under your left arm, against your body.
   (8) While supporting the launcher with your left hand, store the shoulder stop with your right hand.
   (9) Depress and hold the tube release button with your left hand, and rotate the inner tube counterclockwise (against the direction of the yellow arrow) with your right hand.
   (10) Release the tube release button, and collapse the inner tube with your right hand.
   (11) Reinstall the transport safety pin.

M136-SERIES SHOULDER- LAUNCHED MUNITIONS

3-55. To restore M136-series shoulder-launched munitions to a carrying configuration—

   (1) Release the red safety release catch.
   (2) Push forward and to the left on the cocking lever, and let it spring back into the SAFE position.
   (3) Grip the base of the sling on the front of the launcher with your left hand and the shoulder stop with your right hand.
   (4) Remove the munition from your shoulder, rotate your body 180 degrees (keeping the munition pointed downrange), and cradle it in your left arm.
   (5) With the launcher cradled in your left arm, replace the transport safety pin/fork until it is fully seated in the retainer hole.
   (6) Return the rear sight to the battlesight setting (200 meters).

WARNING

Do not close the rear sight cover before ensuring the rear sight is set to 200 meters. This can severely damage the sight.

---

   (7) Fold the rear sight down, hold it down, and close the rear sight cover.
   (8) Fold the front sight down, hold it down, and close the front sight cover.
   (9) Snap the shoulder stop into the closed position.
   (10) *M136A1 Only:* Close the front grip.

M72-SERIES SHOULDER-LAUNCHED MUNITIONS

3-56. To restore M72-series shoulder-launched munitions to a carrying configuration—

CAUTION

After the launcher has been prepared for firing, it is no longer watertight. Therefore, when carrying the launcher, sling it over either shoulder with the muzzle (forward) end down. Only the rocket and rocket motor ignition system are waterproof.

---

(1) Return the trigger arming handle to the SAFE position.
(2) *M72A4/5/6/7:* Return the rear sight to its lowest setting, and ensure the sight peephole is set in the 2-mm day setting.
(3) *M72A4/5/6/7:* Unlock the rear sight by pressing down on the rear sight release tab.
(4) Grip the forward end of the launcher with your left hand and the rear end of the launcher with your right hand.
(5) Remove the munition from your shoulder, rotate your body 90 degrees to the right (keeping the munition pointed downrange), and hold the munition away from your body.

Note. Although the M72-series launcher can be fired from either shoulder, for training purposes, restoring the launcher is performed when fired from your right shoulder.

---

(6) Depress the barrel detent.

WARNING

To prevent injury, remove your thumb from the detent after collapsing the launcher 1/2 to 1 inch.

---

(7) With the firing hand, collapse the launch tube approximately 4 inches.
(8) Fold the rear sight down with the thumb of the firing hand, and hold.

Note. With the improved M72, the rear sight may become locked when the munition is collapsed. If this happens, support the launcher under the nonfiring arm, while keeping the launcher pointed downrange, and then unlock the rear sight with the firing hand.

---

(9) Fold the front sight down with the thumb of the nonfiring hand, and hold.
(10) Partially slide the inner tube into the outer tube until the front sight cover appears in the front opening, ensuring the front sight is under the front sight housing and the rear sight slides inside of the rear sight housing.
(11) Close the launcher completely.
(12) Close the rear cover.
(13) Replace the rear cover transport safety pin.
(14) Secure the sling assembly.
(15) Supporting the munition with your right hand, place the front cover of the sling assembly over the front end of the munition, and secure in place by maintaining pressure on the sling assembly with your right hand.
(16) Holding the sling assembly with your right hand, place the front end of the munition (sling side facing toward your legs) between your left and right boot.

(17) Place your left hand on the rear cover, and hold while pulling up on the sling assembly with your right hand. Place the sling hook on the rear cover.

Note. See the appropriate TM for more information about reinstalling the sling assembly.

ENVIRONMENTAL CONDITIONS

3-57. Shoulder-launched munitions should not be fired when operational temperatures exceed the limits outlined in Table 2-1.

CAUTION

When operating shoulder-launched munitions in cold weather, be aware that bringing a launcher into a warm enclosure may cause damage. The change in temperature will make metal components sweat, and the moisture can cause rust or corrosion.

WARNING

To prevent injury, do not use sharp instruments to chip off snow or ice, or thaw a shoulder-launched munition near a direct flame.

Notes. 1. When operating in rain and snow, protect shoulder-launched munitions from moisture in the same manner as a rifle.

2. For operating and storage limits, see Chapter 2.

DESTRUCTION PROCEDURES (COMBAT ONLY)

3-58. Destruction of any military weapon is authorized only as a last resort to prevent the enemy from capturing or using it. In combat situations, the commander has the authority to destroy weapons, but he must report doing so through the proper channels.

Note. Certain procedures outlined require the use of explosives and incendiary grenades. Related principles and the specific conditions under which destruction occurs are command decisions.

DANGER

BEFORE USING ANY DESTRUCTION PROCEDURE, MOVE TO A SAFE POSITION AND TAKE COVER TO AVOID POSSIBLE INJURY OR DEATH. BEFORE USING DEMOLITIONS FOR ANY REASON, YOU MUST KNOW THE PROPER PROCEDURES OUTLINED IN FM 3-34.214.

METHODS OF DESTRUCTION

3-59. Equipment may be destroyed using several methods. The commander must use his imagination and resourcefulness to select the best method of destruction based on the facilities available. Time is usually critical. Table 3-1 outlines the methods of destruction.

WARNING

These methods are intended as combat expedients only. DO NOT use them for routine disposal operations.

Note. If destruction is directed, appropriate safety precautions must be observed.

DEGREE OF DAMAGE

3-60. The method of destruction used must damage equipment and essential spare parts to the extent that they cannot be restored to usable condition (by repair or by cannibalization) in the combat zone.

DANGERS

1. WHEN USING FIRE TO DESTROY A SHOULDER-LAUNCHED MUNITION, THE TIME REQUIRED TO EXPLODE THE WARHEAD IS UNPREDICTABLE. ALSO, IGNITING THE PROPELLANT CAN CAUSE THE WARHEAD TO FIRE IN ANY DIRECTION, WHICH COULD CAUSE INJURY OR DEATH.

2. OBSERVE THE APPROPRIATE SAFETY PRECAUTIONS WHEN HANDLING DIESEL FUEL. CARELESSNESS COULD CAUSE PAINFUL, EVEN FATAL, BURNS.

3. DO NOT TRY TO USE VEHICLES OR MECHANICAL MEANS TO DESTROY SHOULDER-LAUNCHED MUNITIONS. EITHER METHOD COULD DETONATE THE WARHEAD OR PROPELLANT CHARGE, WHICH COULD CAUSE INJURY OR DEATH.

Table 3-1. Methods of destruction and their applications.

METHOD OF DESTRUCTION	APPLICATION
Demolition	Prepare a 113-gram (1/4 pound) demolition charge. Tape or tie the charge: • M141 BDM – just forward of the firing mechanism. If the launcher is extended, place the charge just behind the shoulder stop. • M136-series shoulder-launched munitions—near the rear of the launcher just forward of the venturi (M136 AT4) or forward of the rear bumper (M136A1 AT4CS) • M72-series shoulder-launched munitions—near the trigger arming handle Dual-prime the charge to reduce the chance of a misfire.
Burning	Construct a pit or trench deep enough to allow 0.6 meters (2 feet) of space between the munitions and the top surface of the ground. Place combustible material, such as wood, paper, or rags, in the pit and then place the munition inside, pointed into the side of the pit and directed away from all friendly Soldiers. Pour diesel fuel or oil over the munitions and the combustible material. Ignite with a combustible train or other safe means. **Note.** Construct a combustible train from any slow-burning material which will allow sufficient time for personnel to take cover before the pit becomes ignited.
Disposal	Shoulder-launched munitions may be disposed of by burying or dumping into a stream or river; however, these methods do not render shoulder-launched munitions unserviceable.
Firing	The easiest and quickest way to destroy a small quantity of shoulder-launched munitions is to fire them. Before using this method, observe all appropriate safety requirements.

DECONTAMINATION PROCEDURES

3-61. The Soldier can use his M291 skin decontamination kit (SDK) or the M295 kit to decontaminate individual equipment.

> *Note.* FM 3-11.5 provides more information about decontamination procedures for personnel, Soldier equipment, and weapons.

> **DANGER**
>
> NEVER USE DECONTAMINATION SOLUTION 2 (DS2) TO DECONTAMINATE ANY SHOULDER-LAUNCHED MUNITION. DS2 CAN DISSOLVE RUBBER AND PLASTIC SEALS.

3-62. The M100 decontamination kit (Figure 3-53) contains a reactive, neutralizing absorbent powder that replaces the DS2 in the M11/M13 decontaminating apparatus, portable (DAP). This powder is used in operator wipe-down (immediate decontamination).

VULNERABLE/SENSITIVE EQUIPMENT

3-63. Some equipment is extremely vulnerable to damage when subjected to decontamination. Most military equipment has not been critically assessed for its ability to withstand decontamination without adverse effects. As more materiel testing is done, specific decontamination instructions will be included in the applicable TMs for all types of equipment.

Optics

3-64. Optical systems are extremely vulnerable to decontamination materials that might scratch or adversely affect the lenses. To clean the optics, wipe them with a soft, nonabrasive material, such as a lens-cleaning tissue, cotton wadding, or a soft cloth dipped in hot, soapy water.

> *Note.* Hot, soapy water is the preferred decontaminant for chemical and biological contamination. The M291 SDK may be used if hot, soapy water is not available.

> **WARNING**
>
> Wipe the optical system with decontaminants. Do not immerse it.
>
> Radiological contamination should be blown off with a stream of air or wiped off with hot, soapy water. Rinse the surface by wiping with a sponge dipped in clean water.
>
> Do not use the M295 individual equipment decontamination kit (IEDK). It contains an abrasive sorbent, which may damage the optics.

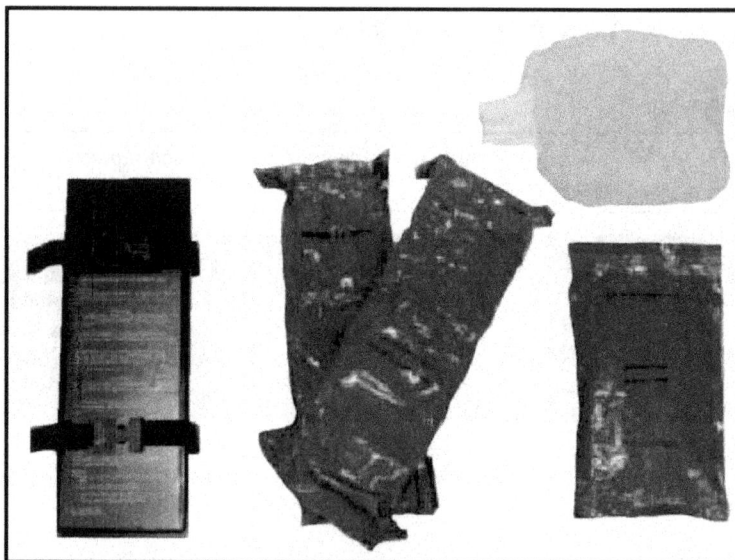

Figure 3-53. M100 decontamination kit.

Ammunition

3-65. To clean contaminated ammunition, apply cool, soapy water with power-driven decontamination equipment (PDDE), brushes, mops, rags, or brooms.

> *Note.* Cool, soapy water is the preferred decontaminant for all types of contamination on ammunition.

CAUTION

Do not use supertropical bleach (STB) or corrosive, nonstandard decontaminants on ammunition. They remove critical markings from ammunition, and may corrode and render ammunition unserviceable.

Chapter 4

MARKSMANSHIP FUNDAMENTALS

The techniques, procedures, and marksmanship skills that enable a Soldier to engage a target with a rifle are also applied for shoulder-launched munitions. The four fundamentals of marksmanship—steady position, aiming, breath control, and trigger squeeze—are just as important when firing shoulder-launched munitions. Although accuracy is not as essential due to the nature of the munition, Soldiers must be able to successfully engage both stationary and moving targets.

SECTION I. FIRING POSITIONS

4-1. Shoulder-launched munitions can be fired from four basic firing positions: standing, kneeling, sitting, and prone. Situation, terrain, and individual preference should govern the selection of the best firing position. Basic safety considerations are the same for all shoulder-launched munitions.

Notes.   1.   Individual physique determines exact body and hand positions. Instructions are given for right-handed firers.

2.   For teaching and demonstration purposes, each position is performed in the open. In combat, Soldiers should fire over or around protective cover.

2.   Soldiers should be trained on the four firing positions, but only the standing and kneeling firing positions are recommended due to overpressure effects at lower ground levels. Low- and ground-level firings pose an increased risk of blast overpressure being deflected onto the firer.

3.   Shoulder-launched munitions are different in design, but they are deployed the same. Because of this, the M136A1 AT4CS will be used to show each firing position.

4.   See Table 2-3 for information about firing limitations.

5.   See Chapter 2 for more information about safety danger zones.

> ## DANGER
> **SHOULDER-LAUNCHED MUNITIONS MUST NOT BE FIRED OVER THE HEADS OF FRIENDLY SOLDIERS.**
>
> **DO NOT FIRE THE LAUNCHER UNTIL THE BACKBLAST AREAS ARE CLEAR OF PERSONNEL, EQUIPMENT, AND OBSTRUCTIONS.**
>
> **DO NOT EXTEND OR ARM THE LAUNCHER UNTIL READY TO FIRE.**
>
> **IN TRAINING, DO NOT ALLOW ANYONE TO ENTER THE AREA BEHIND THE FIRING LINE, OR FORWARD OF THE REAR SAFETY LINE.**

STANDING POSITION

4-2.   Two standing positions are used: a basic standing position and one modified for the Infantry fighting position (for use in combat only).

BASIC STANDING POSITION

4-3.   Unless the Soldier is positioned behind a protective barrier, such as a wall, the basic standing position exposes him to enemy observation and possible suppression more than any other position. To fire from the basic standing position (Figure 4-1)—

WARNING

Always keep the launcher pointed in the direction of fire.

Note. The performance steps start with the shoulder-launched munition on the firing shoulder.

(1)   Spread your feet a comfortable distance apart.
(2)   Move your left foot 15 to 24 inches forward, keeping your hips level, your weight balanced on both feet, and both feet flat on the ground.
(3)   Tuck both elbows tightly into your body.
(4)   *M136A1 AT4CSs:* Hold the front grip firmly with your left hand.
   M72-Series Munitions: Cup the bottom of the launcher.
   M136 AT4s: Hold the forward strap with your left hand.
   M141 BDMs: Slide your left arm inside the sling until it is behind your bicep. Then, grasp the launcher in your left hand, and with your palm up, slide your hand along the tube until the sling tightens and the shoulder stop is seated firmly against your right shoulder. Do not wrap the sling around your left arm, as you would with a rifle (Figure 4-9).
(5)   Place your right hand on the firing mechanism.
   M72-Series Munitions Only: Place your right or left hand on the firing mechanism.
(6)   Place your firing eye 2 1/2 to 3 inches from the rear sight.
   M72A2/3s Only: Place your firing eye at easy reading distance from the rear sight.

Note. To smoothly track a moving target, turn your body at the waist—not with your legs.

Figure 4-1. Basic standing firing position.

MODIFIED STANDING POSITION (FOR USE IN COMBAT ONLY)

> **DANGER**
>
> FIRE M136-SERIES MUNITIONS, M72-SERIES MUNITIONS, OR M141 BDMS FROM THE FIGHTING POSITION WHEN IN COMBAT ONLY. DO NOT FIRE THESE MUNITIONS FROM THE FIGHTING POSITION DURING TRAINING DUE TO THE RISK OF INJURY TO THE OPERATOR.
>
> DO NOT FIRE THE M136 AT4, THE M72, OR THE M141 BDM FROM AN ENCLOSURE.
>
> DO NOT FIRE M136-SERIES MUNITIONS, M72-SERIES MUNITIONS, OR M141 BDMS FROM IN FRONT OF A BARRIER.
>
> TO INCREASE ACCURACY AND REDUCE THE DANGER TO FRIENDLY SOLDIERS, ENSURE THE AREA TO THE REAR OF THE FIRING POSITION HAS NO WALLS, LARGE TREES, OR OTHER OBSTRUCTIONS WITHIN 5 METERS (5 1/2 YARDS). OBSTRUCTIONS DEFLECT BACKBLAST ONTO THE FIRER OR INTO THE FIRING POSITION, INJURING OR KILLING THE FIRER AND ANY OTHER SOLDIER OCCUPYING THE POSITION.
>
> IN THE TWO-SOLDIER INFANTRY FIGHTING POSITION, NONFIRING PERSONNEL MUST REMAIN CLEAR OF THE BACKBLAST AREA. A MODIFIED FIRING POSITION MAY BE CONSTRUCTED TO THE SIDE OF THE TWO-SOLDIER FIGHTING POSITION. FIRING FROM A MODIFIED POSITION REDUCES THE POSSIBILITY OF INJURY TO THE FIRER OR THE OTHER SOLDIER IN THE FIGHTING POSITION, WHILE STILL OFFERING THE FIRER PROTECTION FROM DIRECT ENEMY RETURN FIRE.

> **CAUTION**
>
> Leaders must ensure that shoulder-launched munitions are positioned so that the backblast misses other fighting positions.

4-4. M136-series munitions, M141 BDMs, and M72-series munitions can be fired from the standard Infantry fighting position; however, this firing position is restricted to combat use only. To fire from the modified standing position (Figure 4-2)—

> **DANGER**
>
> WHEN FIRING FROM THE MODIFIED STANDING POSITION, KEEP YOUR BACK AGAINST THE WALL OF THE DUG-IN FIGHTING POSITION TO MINIMIZE DEFLECTION. RAISING THE FRONT END OF THE LAUNCHER CAN CAUSE THE BACKBLAST TO BE DEFLECTED ONTO THE FIRER, CAUSING INJURY OR DEATH. IF THE FIGHTING POSITION RESTRICTS THE BACKBLAST AREA, THE FIRER SHOULD MOVE TO AN ABOVE-GROUND POSITION BEFORE FIRING THE SHOULDER-LAUNCHED MUNITION.

Note. It is much safer to fire a shoulder-launched munition from a hasty fighting position than it is to fire one from the standard Infantry fighting position. Within seconds, Soldiers can fire from the kneeling position and return to cover.

(1) Assume the basic standing position, but instead of stepping forward, lean against the back wall of the fighting position. Do not support your elbows.

(2) Ensure that the rear of the launcher extends beyond the rear of the fighting position.

(3) Ensure that *NONE* of the following are in your backblast area:
- Other Soldiers.
- Other fighting positions.
- Equipment.
- Any part of your own fighting position.
- Obstructions within 5 meters.

Figure 4-2. Modified standing firing position.

KNEELING POSITION

4-5. Two types of kneeling positions are used: basic and modified. The kneeling position offers more stability for engaging targets at longer ranges. It also offers a smaller profile for firing over and from the sides of cover, and from inside urban structures.

BASIC KNEELING POSITION

4-6. The basic kneeling position is the best position for tracking moving targets. Though it is not a supported position, it should be a firm and stable one. To fire from the basic kneeling position (Figure 4-3)—

WARNING

Always keep the launcher pointed in the direction of fire.

---

(1) Begin in basic standing position. Kneel onto your right knee, keeping your left thigh parallel to the ground.

(2) Rotate your lower right leg 90 degrees to the left. This removes your right foot from exposure to the backblast.

(3) Keep your right thigh and back straight and perpendicular to the ground.

(4) Point your left foot in the direction of fire.

(5) Tuck both elbows tightly into your body.

(6) *M136A1 AT4CSs:* Hold the front grip firmly with your left hand.

M72-Series Munitions: Cup the bottom of the launcher.

M136 AT4s: Hold the forward strap with your left hand.

M141 BDMs: Use the weapon sling to increase firer control, as you would with a conventional rifle. Slide your left arm inside the sling until it is behind your bicep. Keep the sling strap behind your bicep; grasp the launcher in your left hand, palm up; and slide your hand along the tube until slack in the sling is taken up. Do not wrap the sling around your left arm as one would with a rifle.

(7) Place your right hand on the firing mechanism.

M72-Series Munitions Only: Place your right or left hand on the firing mechanism.

(8) Place your firing eye 2 1/2 to 3 inches from the rear sight.

M72A2/3s Only: Place your firing eye at easy reading distance from the rear sight.

Note. To smoothly track a moving target, turn your body at the waist—not with your legs.

Figure 4-3. Basic kneeling firing position.

MODIFIED KNEELING POSITION

4-7.   The modified kneeling position is best for engaging stationary targets, since it is a supported position. To fire from the modified kneeling position (Figure 4-4)—

WARNING

Always keep the launcher pointed in the direction of fire.

---

(1) Begin in the basic kneeling position, and sit back on your right heel.

(2) Place the back of your upper left arm on your left knee, making sure you do not have bone-to-bone contact between your left elbow and left knee.

(3) Keep your right elbow tucked in close to your right side.

(4) Use any protective barriers available.

Figure 4-4. Modified kneeling firing position.

SITTING POSITION (FOR USE IN COMBAT ONLY)

4-8.   The sitting position is the most stable firing position. In this position, the Soldier places his arms on his legs for support. Depending on his physique, the firer can use the basic sitting position or the modified sitting position, both of which are suitable for engaging stationary targets.

DANGER

WHEN FIRING FROM THE SITTING POSITION, KEEP THE LAUNCHER PARALLEL TO THE GROUND. RAISING OR LOWERING THE FRONT END OF THE LAUNCHER CAN CAUSE THE BACKBLAST TO BE DEFLECTED ONTO THE FIRER, CAUSING INJURY OR DEATH.

BASIC SITTING POSITION

4-9.   To fire the M136 AT4 or M72 from the basic sitting position (Figure 4-5)—

Note. The M136 AT4 is used to show proper sitting fire techniques that can be used for firing M72 shoulder-launched munitions.

> ### WARNING
>
> **The M41 BDM and M136A1 AT4CS should not be fired from the sitting position in accordance with the M41 BDM and M136A1 AT4CS TMs.**
>
> **Always keep the launcher pointed in the direction of fire.**

(1)   Sit on your buttocks, facing the target.

(2)   Spread your feet a comfortable distance apart.

(3)   Lean forward, and place the backs of your upper arms on your knees (avoiding bone-to-bone contact) or place your elbows inside of your thighs.

(4)   Hold the forward strap firmly with your left hand.

   M72-Series Munitions: Cup the bottom of the launcher.

(5)   Place your right hand on the firing mechanism.

   M72-Series Munitions Only: Place your right or left hand on the firing mechanism.

(6)   Place your firing eye 2 1/2 to 3 inches from the rear sight.

   M72A2/3s Only: Place your firing eye at easy reading distance from the rear sight.

Figure 4-5. Basic sitting firing position.

MODIFIED SITTING POSITION

4-10. To fire from the modified sitting position (Figure 4-6)—

Note. See Table 2-3 for more information about firing limitations.

WARNING

The M41 BDM and M136A1 AT4CS should not be fired from the sitting position in accordance with the M41 BDM and M136A1 AT4CS TMs.

Always keep the launcher pointed in the direction of fire.

---

(1) Begin in the basic sitting position.
(2) Cross your ankles for added support.
(3) Raise or lower your knees to adjust for elevation on the target.

Figure 4-6. Modified sitting firing position.

PRONE POSITION (FOR USE IN COMBAT ONLY)

4-11. The prone position is the least stable position and, due to its proximity to the ground, is the most dangerous position in regards to potential backblast injury. However, it also offers the most protection from enemy observation. Ideally, the ground should slope downward from the rear of the launcher, which reduces the effects of the backblast. To fire from the prone position (Figure 4-7)—

WARNING

Always keep the launcher pointed in the direction of fire.

DANGER

FIRE SHOULDER-LAUNCHED MUNITIONS FROM THE PRONE POSITION WHEN IN COMBAT ONLY. DO NOT FIRE SHOULDER-LAUNCHED MUNITIONS FROM THE PRONE POSITION DURING TRAINING DUE TO THE RISK OF INJURY TO THE OPERATOR.

MAINTAIN A 90-DEGREE (WHEN FIRING AN M136 AT4), 45-DEGREE (WHEN FIRING AN M136A1 AT4CS OR M141 BDM), 35-DEGREE (WHEN FIRING AN IMPROVED M72), OR 30-DEGREE (WHEN FIRING AN M72A2/A3) ANGLE TO THE DIRECTION OF FIRE. FAILURE TO DO SO COULD CAUSE INJURY TO THE FIRER.

---

(1) *M136 AT4s:* Lie on your stomach, with your body at a 90-degree angle to the direction of fire.
M136A1 AT4CSs or M141 BDMs: Lie on your stomach, with your body at a 45-degree angle to the direction of fire.
Improved M72s: Lie on your stomach, with your body at a 35-degree angle to the direction of fire.
M72A2/A3s: Lie on your stomach, with your body at a 30-degree angle to the direction of fire.

(2) Place your body and legs to the left of the direction of fire. Place your right leg (firing side) over your left leg. Ensure that neither your body nor your legs are in the backblast area.

(3) Hold the launcher in place against your upper right arm.

Note. Unlike other firing positions, the prone position prevents you from placing the launcher on your right shoulder.

---

(4) *M136A1 AT4CSs:* Hold the front grip firmly with your left hand, while keeping the launcher in place against your upper right arm.
M141 BDMs or M72-Series Munitions: Cup the bottom of the launcher.
M136 AT4s: Hold the forward strap with your left hand.

(5) Place your right hand on the firing mechanism.
M72-Series Munitions Only: Place your right or left hand on the firing mechanism.

Note. For stability, apply extra pressure on the firing mechanism with your firing hand.

---

(6) Place your firing eye 2 1/2 to 3 inches from the rear sight.
M72A2/3s Only: Place your firing eye at easy reading distance from the rear sight.

Figure 4-7. Prone firing position.

SECTION II. TARGET ENGAGEMENT PROCEDURES

4-12. Many factors contribute to shoulder-launched munition marksmanship. These factors are grouped into six basic areas: steady hold, range estimation, speed estimation, aiming procedures, breath control, and trigger manipulation. All of these areas are applied to the integrated act of firing.

Notes.   1.   Except for those on M72A2/A3 launchers, shoulder-launched munition sights are similar in design, so aiming techniques are basically the same. For this reason, the M136A1 AT4CS is used to demonstrate and explain target engagement procedures.

2.   M72-series launchers can be fired with your right or left hand. The trigger arming handle and trigger is positioned on the top of the launcher, whereas other arming devices and triggers are located on the right side of the launcher. Instructions are given for right-handed firers.

STEADY HOLD

4-13. Maintaining a steady hold involves holding the launcher as steady as possible while sighting and firing. To maintain the proper sight picture and sight alignment (Figure 4-8)—

(1)   Hold the launcher in a tight, comfortable position so that it becomes a natural extension of your body.

(2)   Keep your elbows close to your body to help balance the launcher and prevent you from jerking or flinching when you fire.

(3)   *M136A1 AT4CSs:* Grasp the front grip (the sling, if the front grip has been damaged) with your left hand, and pull back on the launcher to seat the shoulder stop firmly against your right shoulder. Ensure your right arm, including your elbow, is through the carrying sling.

M141 BDMs: Slide your left arm inside the sling until it is behind your bicep. Then, grasp the launcher in your left hand, and with your palm up, slide your hand along the tube until the sling tightens and the shoulder stop is seated firmly against your right shoulder. Do not wrap the sling around your left arm, as you would with a rifle (Figure 4-9).

M136 AT4s: Grasp the carrying sling where it attaches to the launcher near the muzzle with your left hand, and pull back on the launcher to seat the shoulder stop firmly against your right shoulder.

M72-series Munitions: Place your nonfiring hand under the launcher, between the safety handle and front sight, and pull back on the launcher to seat the shoulder stop/rear cover firmly against your firing shoulder.

Notes.   1.   M72-series launchers do not have carrying slings.

            2.   M72-series munitions can be fired with your left or right hand.

(5) Place your firing hand on the trigger mechanism.

Note. Firing from a supported position increases accuracy and improves the odds for a first-round hit or kill on stationary targets. However, supported positions can restrict movement when engaging moving targets.

Figure 4-8. Steady hold position—M136A1 AT4 confined space.

Figure 4-9. Steady hold position—M141 bunker defeat munition.

RANGE ESTIMATION

4-14. Shoulder-launched munitions come equipped with fixed sights that adjust to a determined range-to-target from the minimum engagement range to the maximum effective range of the munition.

Note. M72A2/A3 fixed sights are not adjustable.

4-15. Methods of estimating range include (listed from the most to the least accurate)—
- Using handheld laser rangefinders.
- Using pair and sequence methods of target engagement.
- Estimating range visually.

Notes.   1.   See FM 3-25.26 for more information about range estimation.

           2.   Do not use the stadia lines on the front sight of the M72A2/A3 to estimate range, because they are inaccurate. The sights are not up-to-date with changes made to modern enemy armored vehicle dimensions.

SPEED ESTIMATION

4-16. To estimate the speed of a moving vehicle, Soldiers should estimate how far the vehicle travels in one second. Figure 4-10 depicts speed estimation for a target in a crossing pattern. This technique works when an object is available for gauging target speed; however, when engaging targets in open terrain using a crossing pattern, Soldiers should engage targets using pair or volley fire.

Note. See Chapter 5 for more information about methods of engagement.

4-17. To determine target speed in wooded or urban terrain—

4-18. Start when the front end of the vehicle passes the object.

     (1)   Count one second (the phrase "one thousand and one" takes about one second).

     (2)   Observe how much of the vehicle passes the object:

- If more than half of the vehicle passes the object, estimate it as a fast-moving vehicle (10 miles per hour or faster).
- If less than half of the vehicle passes the object, estimate it as a slow-moving vehicle (less than 10 miles per hour).

4-19. Once firers learn to estimate speeds and engage moving targets at known ranges, they should rehearse until they achieve a high hit-to-kill ratio. As their abilities improve, unit leaders should vary the ranges, speeds, and types of armored vehicles used.

Note. Speed estimation is not required for engaging a target moving away from the firer (rear silhouette) and coming toward the firer (frontal silhouette), when the target is within the munition's maximum effective range.

Figure 4-10. Speed estimation.

AIMING PROCEDURES

4-20. Aiming procedures include placing the eye correctly, obtaining a sight picture, and aligning the sight.

EYE RELIEF

4-21. To achieve correct eye relief, place your firing eye (Figure 4-11)—

- *M136-series Munitions and M141 BDMs:* Between 2 1/2 to 3 inches from the rear sight. The white semicircle on the front sight will match the curve of the rear sight peep when you are in the correct position.
- *Improved M72s:* Between 2 1/2 to 3 inches from the rear sight. The left and right lead posts on the front sight blade are just visible in the rear sight aperture when you are in the correct position.
- *M72A2/A3s:* At an easy reading distance from the rear sight. The front sight blade is just visible in the rear sight aperture when you are in the correct position.

WARNING

DO NOT place your eye closer to the rear sight than the distance indicated. The munition's recoil could cause the rear sight to injure your firing eye.

Figure 4-11. Eye placement for shoulder-launched munitions.

SIGHT PICTURE

4-22. To achieve a proper sight picture—

(1) Position the front sight on the target.
(2) Adjust the rear sight for the correct range.
(3) Sight the munition.
(4) Place the front sightpost on the target as appropriate for the desired point of aim or the target's speed and direction of movement (Tables 4-1, 4-2, and 4-3).

Notes. 1. ***M136-Series Munitions and the M141 BDMs:*** You have achieved a correct sight alignment when the white semicircle (half-moon) at the bottom of the front sightpost can be seen.

2. ***M72-Series Munitions:*** You have achieved a correct sight alignment when the left and right lead posts (improved M72) or the range index markings (M72A2/A3) on the front sight blade are just visible in the rear sight aperture.

**Table 4-1. Sight picture for various targets for
M136-series shoulder-launched munitions and M141 bunker defeat munitions.**

TYPE OF TARGET	SPEED AND DIRECTION OF MOVEMENT	PROCEDURE FOR PROPER SIGHT PICTURE
Stationary Targets 	Fixed positions and fortifications, as well as vehicles moving directly toward or away from the firer	Place the center sightpost in the center of the target or at the desired hit point.
Slow-Moving Targets 	Targets with an estimated speed of 10 miles per hour or less or those moving in an oblique direction	Place the sightpost on the front or leading edge of the vehicle.
Fast-Moving Targets 	Targets estimated to be moving faster than 10 miles per hour	Place the left or right top of the white half-moon on the center of the target.

Table 4-2. Sight picture for various targets for improved M72.

TYPE OF TARGET	SPEED AND DIRECTION OF MOVEMENT	PROCEDURE FOR PROPER SIGHT PICTURE
Stationary Targets 	Fixed positions and fortifications, as well as vehicles moving directly toward or away from the firer	Place the center sightpost in the center of the target or at the desired hit point.
Slow-Moving Targets 	Targets with an estimated speed of 15 miles per hour or less	Place the sightpost on the front or leading edge of the vehicle.
Fast-Moving Targets 	Targets estimated to be moving faster than 15 miles per hour	Place the top of the lead post on the center of the target.

Table 4-3. Sight picture for various targets for M72A2/A3.

TYPE OF TARGET		SPEED AND DIRECTION OF MOVEMENT	PROCEDURE FOR PROPER SIGHT PICTURE
Stationary Targets FRONTAL	FLANK	Fixed positions and fortifications, as well as vehicles moving directly toward or away from the firer	Position the front sight on the target. Place the correct vertical range line in the center of the target.
Slow-Moving Targets LEFT SIDE	RIGHT SIDE	Targets with an estimated speed of 5 miles per hour or less or those moving in an oblique direction	Place the left or right lead cross mark on the vehicle's center of mass.
Fast-Moving Targets FLANKING LEFT SIDE	RIGHT SIDE	Targets estimated to be moving faster than 5 miles per hour	Place the left or right lead cross mark on the leading edge of the vehicle.

BREATH CONTROL

4-23. Breath control is as important when firing a shoulder-launched munition as it is when firing an individual weapon. Improper breath control while firing can cause a miss. To control breathing—

(1) Breathe deeply a couple of times.

(2) Take one last deep breath.

(3) Exhale partly.

(4) Hold your breath.

(5) Sight the munition.

(6) Fire the munition.

Note. This technique is also used in rifle marksmanship. See Chapter 4 of FM 3-22.9 for more information.

TRIGGER MANIPULATION

4-24. Trigger manipulation differs according to the munition being fired:

- *M141 BDMs (Figure 4-12) or M136-series Munitions (Figure 4-13):* Apply firm and steady forward pressure to the trigger with the thumb of your firing hand.

- *M72-series Munitions (Figure 4-14):* Place the thumb of your right hand under the launcher, and your fingertips on the trigger, while keeping the palm of your hand firmly against the side of the launch tube.

Notes. 1. Soldiers can practice trigger manipulation and control techniques on an expended launcher or a FHT.

2. Appendix A of FM 3-22.9 provides a technique (dime/washer exercise) for teaching rifle trigger squeeze discipline. This exercise can be applied to shoulder-launched munition training.

Figure 4-12. Trigger manipulation of a M141 bunker defeat munition.

Figure 4-13. Trigger manipulation of a M136-series shoulder-launched munition.

Figure 4-14. Trigger manipulation of a M72-series shoulder-launched munition.

INTEGRATED ACT OF FIRING

4-25. To properly perform the integrated act of firing, the firer focuses on the front sight to obtain correct sight alignment, and then places the aimpoint to complete the sight picture. He shifts or adjusts the position of the launcher as necessary. The firer maintains the sight picture as he presses the trigger.

Note. The EST 2000 can be used to test learned firing skills and select designated marksmen.

Chapter 5

EMPLOYMENT CONSIDERATIONS

This chapter discusses employment considerations for shoulder-launched munitions. All techniques require at least basic marksmanship skills; those that require advanced skills are identified.

METHODS OF ENGAGEMENT

5-1.   The leader evaluates the situation on the ground to determine which method of engagement to use. The four engagement methods include single, sequence, pair, and volley firing.

5-2.   Communication is required to use all of these methods. Leaders control all unit fire and communicate this information to the entire unit in accordance with the unit SOP. To properly engage a target, leaders should inform shoulder-launched munitions designated marksmen of the following:

- Use of the proper munition for a given target.
- Target priority.
- Target engagement areas.
- Method of engagement.
- Range and lead to target (if known).
- Command or signal to fire.
- Command or signal to cease fire.

SINGLE FIRING

5-3.   Single firing involves using a single Soldier armed with a single shoulder-launched munition to engage a target. This method requires the Soldier to hit a vital part of the target to get the desired effect.

SEQUENCE FIRING

5-4.   Sequence firing involves using a single firer equipped with two or more shoulder-launched munitions (prepared for firing) to engage a target. After engaging with the first round and observing the impact, the firer engages with additional rounds until he disables or destroys the target or runs out of rounds.

Notes.   1.   Sequence firing is not the recommended means of engaging any type of target, day or night. The firer is more susceptible to receiving direct and indirect enemy fire when firing multiple shots from one location.

2.   Shoulder-launched munitions do not have dedicated NVDs; firers must take those devices from other weapons. Sequence firings require the firer to remove the devices after each shot and then reattach them to another round of munition. Leaders are encouraged to use pair and volley fires or other means to illuminate targets when conducting operations during limited visibility conditions.

PAIR FIRING

5-5.   Pair firing involves using two or more firers equipped with shoulder-launched munitions (prepared for firing) to engage a single target. The first firer informs the others of the distance to the target, and if it is moving, the estimated speed. If the impact of his round proves his estimate to be correct, the other firer(s) engage(s) the target. If the impact of the round proves his estimate to be incorrect, the second firer informs the others of his estimate and engages the target.

5-6.   Firing in pairs is the recommended method for creating man-sized holes in structural walls using a M141 BDM. For example: One firer creates the initial entry point, and the other firer places a round in close proximity to widen the entry point.

VOLLEY FIRING

5-7.   Volley firing involves using two or more firers to engage a single target (when the range is known) at the same time on a prearranged signal.

5-8.   Volley firing can be the most effective means of engagement, as it places the most possible rounds on one target at one time, increasing the possibility of a kill. Volley firing is the recommended method of engaging more than one target or target area, day or night.

> *Note.* When conducting operations during limited visibility conditions, NVSs should only be needed for the initial engagement, which eliminates the need to detach and reattach these devices. After the initial assault, the commander may use other sources to see the operational environment.

ENGAGEMENT OF VEHICLES

5-9.   There are two types of vehicles:
- Armored vehicles.
- Non-armored vehicles.

5-10. Although current shoulder-launched munitions can be used against armored and non-armored vehicles, matching the right munition to the target can mean success or failure. Table 5-1 compares the effects of different munitions on vehicle types.

5-11. Armored vehicle kills are classified according to the level of damage achieved (Table 5-2). Table 5-1 classifies degrees of damage using these levels.

> *Note.* Firers should always aim at the vehicle's center of mass to increase the probability of a hit.

Table 5-1. Effects of different munitions on vehicle types.

MUNITION	EFFECTS	REMARKS
Heavy-Armored Vehicle		
The older the vehicle model, the less protection it has against shoulder-launched munitions. Newer versions may use bolt-on (appliqué) armor to improve their survivability. Some vehicles are equipped with reactive armor, which consists of metal plates and plastic explosives.		
M141 BDM	Can cause a mobility kill by disabling the vehicle's suspension system	The M141 BDM should be a last resort when engaging armored vehicles.
M136-series	Causes only a small entry hole, though some fragmentation or spallation may occur	Reactive armor usually covers the front and sides of the vehicle, and can defeat shaped-charge weapons; however, the munition can restrict the vehicle's mobility and may destroy the vehicle if the round hits a vulnerable spot, such as the engine compartment area.
M72-series	Causes only a small entry hole, though some fragmentation or spallation may occur	
Light-Armored Vehicle		
All current shoulder-launched munitions are capable of destroying most light-armored vehicles, if the round hits a vulnerable spot, such as the engine compartment area, or fuel tank. Unit leaders should provide squad and platoon supporting fires when engaging light-armored troop carriers. Any infantry troops that survive the initial assault may dismount and return fire.		
M141 BDM	Can cause a catastrophic kill, if the round hits a vulnerable spot, such as the engine compartment area or fuel tank	N/A
M136-series	Can cause a catastrophic kill, if the round hits a vulnerable spot, such as the engine compartment area or fuel tank	N/A
M72-series	Can cause a catastrophic kill, if the round hits a vulnerable spot, such as the engine compartment area or fuel tank	N/A
Non-Armored Vehicles		
Non-armored vehicles, such as trucks and cars, are considered soft targets. Firing along their length (flank) offers the greatest chance of a kill, because this type of shot is most likely to hit their engine block or fuel tank. Front and rear angles offer a much smaller target, reducing the chance of a first time hit.		
M141 BDM	Causes a catastrophic kill	
M136-series	May penetrate, but will pass through the body with limited damage unless the rocket hits a vital part of the engine	When engaging enemy-used privately-owned vehicles (POVs) with M136- or the M72-series munitions, do not fire at the main body. Instead, fire at the engine compartment area.
M72-series	May penetrate, but will pass through the body with limited damage unless the rocket hits a vital part of the engine	

Table 5-2. Armored vehicle kills.

TYPE OF KILL	PART OF VEHICLE DAMAGED OR DESTROYED	CAPABILITY AFTER KILL
Mobility Kill	Suspension (track, wheels, or road wheels) or power train (engine or transmission) has been damaged.	Vehicle cannot move, but it can still return fire.
Firepower Kill	Main armament has been disabled.	Vehicle can still move, so it can get away.
Catastrophic Kill	Ammunition or fuel storage section has been hit by more than one round.	Vehicle is completely destroyed.

ENGAGEMENT OF FIELD FORTIFICATIONS AND BUILDINGS

5-12. Shoulder-launched munitions can be used to engage various types of field fortifications and buildings, with different effects for the type of shoulder-launched munition used.

M141 BUNKER DEFEAT MUNITION

5-13. The M141 BDM was designed to better enhance the destruction of field fortifications and buildings. The M141 BDM contains a high-explosive, dual purpose (HEDP) round with a dual-mode fuze that automatically adjusts for the type of target on impact. For soft targets, such as sandbagged bunkers, the M141 BDM warhead automatically adjusts to delayed mode and hits the target with high kinetic energy. This energy propels the warhead through the barrier and into the fortification or building, where the fuze detonates the warhead and causes greater damage.

5-14. Table 5-3 provides recommendations and considerations for M141 BDM use.

Table 5-3. Effects of M141 bunker defeat munitions on field fortifications or bunkers.

AIMPOINT	EFFECT WHEN MUNITION IS FIRED AT AIMPOINT		RECOMMENDED FIRING TECHNIQUE
Bunkers	Rounds fired into firing ports or apertures can destroy standard earth and timber bunkers, and hasty urban fighting positions (i.e., vehicles, metal dumpsters). Rounds will detonate inside the rear of the position, causing major structural damage. Damage to enemy equipment may be minor unless it is hit directly. The round will cause injury or death to occupants.		Coordinate fire: Fire a shoulder-launched munition at and through firing ports.
Buildings	Windows/ Doorways	Rounds fired through windows and doorways can destroy the contents of the building. Destruction may not be contained within a single room. Rounds and/or debris from the round and material may pass through into other sections of the building, causing collateral damage. Damage to enemy equipment may be minor unless it is hit directly. The round will cause injury or death to occupants.	Coordinate fire: Fire a M141 BDM at the center of the visible part of a window or door.
	Walls	Rounds fired at walls will penetrate double-reinforced concrete walls up to 8 inches thick, and triple-brick structures. The initial blast will open a hole in the wall, but may or may not completely penetrate the building.	Coordinate fire: Fire one or more M141 BDMs at the center of the desired location for the opening. Fire a second round through the opening to destroy targets within the structure. *Note.* It takes more than one round to create a man-size hole. Use pair or volley fire, placing the rounds about 12 to 18 inches apart.
Underground Openings	Rounds fired through underground openings can collapse the opening or destroy the contents within it. Destruction may not be contained within the opening. Rounds and/or debris may pass through into other sections of the opening, causing further damage. Damage to enemy equipment may be minor unless it is hit directly. The round will cause injury or death to occupants at the front entrance, and others farther into the opening may be incapacitated or die from the concussion, heat, and debris caused by the explosion.		Coordinate fire: Fire one or more M141 BDM.

M136- AND M72-SERIES MUNITIONS

5-15. M136- and M72-series launchers have high-explosive antitank (HEAT) warheads. These warheads are designed to burn through the material that protects enemy heavy-and light-armored vehicles, and they create spalls upon reentry. These spalls can ignite fuel and ammunition, and injure or kill personnel.

5-16. When used against field fortifications and buildings, however, these munitions have little effect. If the alternatives shown in Table 5-4 are used, Soldiers may be able to gain a temporary advantage.

Table 5-4. Effects of M136- and M72-series munitions on field fortifications or bunkers.

AIMPOINT		EFFECT WHEN MUNITION IS FIRED AT AIMPOINT	RECOMMENDED FIRING TECHNIQUE
Bunkers		Firing at the bunker causes the round to detonate outside the fighting position or inside the bunker, creating only a small hole in the bunker, dust, or minor structural damage to the position, but no damage to personnel or equipment unless they are hit directly.	
Buildings	Windows/ Doorways	The round may travel completely through the structure before detonating. If not, it creates dust and causes minor structural damage to the rear wall, but little damage to personnel or equipment unless they are hit directly.	Coordinate fire: Fire 6 to 12 inches from the sides or bottom of a window. M136- and M72-series rounds explode on contact with brick or concrete, creating an opening with a size determined by the type of round used.
	Walls	The round detonates on contact, creating dust and causing a small hole and minor structural damage, but little damage to personnel or equipment, unless they are hit directly.	
	Corners	Corners are reinforced and, therefore, harder to penetrate than other parts of a wall. The munition will detonate sooner on a corner than on a less dense surface. Detonation should occur in the targeted room, creating dust and causing overpressure, which can temporarily incapacitate personnel inside the structure near the point of detonation. *Note.* M136-series munitions cause more overpressure than M72-series munitions.	
Underground Openings		Rounds fired around underground openings can collapse the opening. Damage to enemy equipment may be minor unless it is hit directly. Heat and debris from the round will cause injury or death to occupants at the front entrance.	Coordinate fire: Fire one or more M136- or M72-series munitions.
Note. Fire small arms at enemy-held positions to prevent personnel within from returning fire.			

ENGAGEMENTS CONDUCTED DURING LIMITED VISIBILITY

5-17. To avoid fratricide, leaders must ensure all designated marksmen are trained for operations conducted during limited visibility conditions. These engagements can be conducted using various NVDs or artificial illumination; however, the use of NVDs or artificial illumination can reduce the ability to see aimpoints clearly and identify targets.

> *Note.* On the M141 BDM, the M136-series munition, and the improved M72 launcher with adjustable sights, use the 7-mm peephole for firing during limited visibility conditions. On the M72A2, use the front sight illuminated range marks at the 100- and 150-meter points to engage targets in low light.

Night Vision Devices

WARNING

To reduce the risk of detection by an enemy wearing night vision goggles (NVGs), avoid prolonged activation of the IR aiming light prior to firing.

The beam of an IR aiming light is more detectable to an enemy using NVGs when shining through smoke, fog, and rain. Avoid prolonged activation of the aiming light in these conditions.

Note. Chapter 1 describes the various NVDs that can be used, and Appendix C gives the mounting and alignment procedures for each NVD.

5-18.   Because shoulder-launched munitions are discarded after firing, the munition has no dedicated sight systems other than those permanently attached to the launcher; therefore, NVDs must be attached to the munition before use. When attaching NVDs, use the following guidelines:

- The M141 BDM has a permanently attached mounting rail that enables mounting of the NVS or the aiming light.
- The M136A1 AT4CS has permanently attached mounting rails that enable simultaneous mounting of the NVS and the aiming light.
- Before a NVD can be used on the M72-series or the M136 AT4, a mounting bracket must be attached. This bracket will only support one device at a time: a NVS or an aiming light.

Notes.   1.   See Appendix C for more information about NVD use.

2.   The M72A4/A5/A6/A7 comes with a forward mounting rail used for mounting aiming lights.

Artificial Illumination

5-19. Illumination can distort the target when placed between the firer and the target. If artificial illumination is used in limited visibility conditions, it should be placed above and slightly beyond the target.

ENGAGEMENT IN CHEMICAL, BIOLOGICAL, RADIOLOGICAL, AND NUCLEAR CONDITIONS

5-20. Wearing a protective mask limits the firer's ability to sight the munition, and wearing chemical, biological, radiological, and nuclear (CBRN) gloves limits his ability to manipulate the firing mechanism.

Sighting the Munition

5-21. To properly sight the munition while wearing the protective gas mask, the firer may have to rotate the launcher slightly counterclockwise. The mask affects depth perception and distorts images, making determining the location, identification, and range-to-target more difficult.

Firing the Munition

5-22. Shoulder-launched munitions practice events provide firing exercises for Soldiers wearing MOPP gear. Soldiers should also practice manipulating the firing mechanism while wearing CBRN gloves.

Note. Before conducting LFXs, Soldiers should practice firing while wearing MOPP gear by using EST 2000 firing exercises.

ENGAGEMENT FROM AN ENCLOSURE

DANGER

THE M136 AT4, M141 BDM, AND M72-SERIES MUNITION MUST NEVER BE FIRED FROM AN ENCLOSURE. THE OVERPRESSURE AND BLAST CAN KILL, SERIOUSLY INJURE, OR DEAFEN THE FIRER AND/OR ANY OTHER PERSONNEL IN THE ENCLOSURE.

THE M136A1 AT4CS MUNITION HAS BEEN RATED SAFE FOR USE FROM AN ENCLOSURE, BUT ONLY WHEN THE ENCLOSURE MEETS THE MINIMUM REQUIREMENTS (LISTED IN THIS SECTION).

5-23. Firing from an enclosure creates unique hazards. As such, leaders must consider several safety factors before firing from enclosures.

Note. The M136A1 AT4CS is the only shoulder-launched munition proven safe for firing from enclosures; however, enclosures must meet the following specifications. See Chapter 2 for more information about using the M136A1 AT4CS to engage targets from enclosures.

CONSTRUCTION

5-24. The building must be sturdily constructed to reduce the structural damage that would occur in a weakly constructed enclosure (such as one made of wood or stucco).

SIZE

5-25. Properly positioning the launcher within the enclosure is vital to the safety and survival of all personnel in the enclosure. The launcher should be positioned so that the firer is as far away as possible from the backblast area. At a minimum, the enclosure should be 12 by 15 feet, with a ceiling height of 7 feet or more to allow for blast overpressure.

VENTILATION

5-26. Without sufficient ventilation, blast overpressure can weaken or collapse the walls. To increase ventilation and reduce overpressure, noise, and blast effects—
- Provide at least 20 square feet of ventilation (such as a standard 3-by 7-foot doorway) directly behind the firer.
- Open or remove all doors and windows.

Note. If a room has only one opening to the rear of the launcher, knock several 3- to 4-foot holes between the wall supports. This will allow more blast overpressure to escape.

WARNING

DO NOT remove wall supports. Doing so can weaken the foundation.

- On the front wall, remove only those portions of the window that may restrict firing the shoulder-launched munition. Muzzle clearance must be at least 4 inches (10 cm).

CAUTION

M136A1 AT4CS rocket fins deploy to a 10-inch diameter. Take care not to impact the window frame when firing the M136A1 AT4CS from an enclosure.

- If a window dressing is present (curtains/blinds), leave it in place until ready to fire. Removal will draw attention to the position.

PROTECTION

5-27. Personnel should reinforce firing positions from the inside to help protect the firer from enemy direct-fire weapons.

OBJECTS AND DEBRIS

5-28. Any objects or debris within the room must be removed so that blast overpressure will not cause them to fly around the room, possibly injuring personnel.

PERSONNEL POSITIONS

5-29. If any other Soldiers are present, they must remain to the side of the shoulder-launched munitions fired. Soldiers in support of shoulder-launched munition firers should avoid standing in corners or near walls. If possible, they should construct reinforced positions that will protect them from the effects of blast overpressure.

Note. See Chapter 2 for more information about firing shoulder-launched munitions from the inside of structures.

WARNING

To avoid injuring the eardrums, Soldiers must wear the approved brand of ear protection. Shoulder-launched munition firers must alert Soldiers in close proximity before firing.

ENGAGEMENT BEYOND MAXIMUM EFFECTIVE RANGE

5-30. A skilled firer can engage targets beyond the munition's maximum effective range. Beyond the maximum effective range, the firer must aim higher than the target's center of mass and apply additional lead for moving targets. Commanders must realize that accuracy is reduced at these ranges. Also, firing at these ranges reveals the firing position to the enemy.

Appendix A

M72 PRACTICE FIRING TABLES

This appendix provides firing tables for M72-series shoulder-launched munitions. The M72AS subcaliber training launcher is used to fire the practice firing tables for M72-series shoulder-launched munitions, and DA Form 7678 (Day and Night Fire-- M72 [M72AS 21-mm Subcaliber Training Launcher], shown in Figure A-1) is used to score them. The tasks, conditions, and standards for the instructional day and night firing tables are provided in Tables A-2 and A-4.

DAY AND NIGHT FIRE
M72 (M72AS 21-MM SUBCALIBER TRAINING LAUNCHER)

For use of this form, see TM 3-23.25. The proponent agency is TRADOC.

NAME _John Doe_		RANK _SPC_		UNIT _B Co 2nd BN/29 Rgt_	
DATE _22 April 2010_		EVALUATOR'S NAME _James Smith_		EVALUATOR'S RANK _SGT_	

TABLE 1—PRACTICE DAY FIRE (STATIONARY)

ROUND	TYPE OF TARGET	RANGE (METERS)	FIRING POSITION	HIT	MISS
1	Stationary	150 to 200	Standing	X	
2	Stationary	200 to 250	Kneeling*	X	
3	Stationary	200 to 250	Modified Kneeling	X	
TOTAL				3	Ø

* The firer wears MOPP when firing from this position.

TABLE 2— PRACTICE DAY FIRE (MOVING)

ROUND	TYPE OF TARGET	RANGE (METERS)	FIRING POSITION	HIT	MISS
1	Moving	100 to 150	Standing	X	
2	Moving	150 to 200	Kneeling	X	
TOTAL				2	Ø

TABLE 3— PRACTICE NIGHT FIRE (STATIONARY/MOVING)

ROUND	TYPE OF TARGET	RANGE (METERS)	FIRING POSITION	HIT	MISS
1	Stationary	200 to 250	Standing	X	
2	Moving	150 to 200	Kneeling		X
TOTAL				1	1

PRACTICE FIRE SCORE			QUALIFICATION SCORE RATINGS	
TABLE	HIT	MISS		
1	3	Ø		
2	2	Ø		
3	1			

THE FIRER WILL BE ISSUED 7 ROUNDS: 5 ROUNDS FOR THE DAY PRACTICE FIRING TABLES AND 2 ROUNDS FOR THE NIGHT PRACTICE FIRING TABLES.

FIRE ALL ENGAGEMENTS IN SEQUENCE.

DO NOT ANNOUNCE THE RANGES TO TARGETS.

AT RANGES BEYOND 250 METERS, SOLDIERS WILL NOT BE ABLE TO OBSERVE TRACER IMPACT. THIS ISSUE IS RESOLVED WHEN USING AN MPRC, AS TARGETS MOVE DOWN UPON IMPACT.

☐ 7 OF 7 HITS — EXPERT

☒ 6 OF 7 HITS — 1st CLASS

☐ 5 OF 7 HITS — 2nd CLASS

☐ 4 AND BELOW — UNQUALIFIED

FIRER'S SIGNATURE _John Doe_	DATE _4/22/10_	SCORER'S INITIALS _JS_	DATE INITIALED _4/22/2010_
RANGE OIC'S SIGNATURE _Henry Moore_		RANK _SFC_	DATE _4/22/2010_

DA Form 7678, OCT 2010

APD PE v1.00ES

Figure A-1. Example of completed DA Form 7678
(Day and Night Fire—M72 [M72AS 21-mm Subcaliber Training Launcher]).

Note. A blank copy of the form is located at the end of this publication for local reproduction on 8 1/2- by 11-inch paper.

PRACTICE DAY FIRE

A-1.   Soldiers conduct practice day fire using a M72AS subcaliber training launcher.

Note. During the execution of this training, range safety personnel should load the subcaliber training launcher and perform any necessary maintenance.

A-2.   Table A-1 shows the distribution of rounds.

Table A-1. Distribution of rounds for practice day fire.

ROUND	TYPE OF TARGET	RANGE (METERS)	FIRING POSITION
Practice Stationary (M72AS Subcaliber Training Launcher)			
1	Stationary	150 to 200	Standing
2	Stationary	200 to 250	Kneeling*
3	Stationary	200 to 250	Modified Kneeling
Practice Moving (M72AS Subcaliber Training Launcher)			
1	Moving	100 to 150	Standing
2	Moving	150 to 200	Kneeling
* The firer wears MOPP gear when firing from this position.			

A-3.   Soldiers fire five rounds using a M72AS subcaliber training launcher: three rounds at stationary targets at ranges of 150 to 300 meters and two rounds at moving targets at ranges of 100 to 200 meters. The purpose of this firing is to determine the firer's ability to estimate range to the target during day conditions, demonstrate correct firing positions, apply the fundamentals of marksmanship, and achieve accuracy while receiving blast overpressure effects. Table A-2 shows the task, conditions, and standards for this training.

Note. Soldier accuracy deteriorates after experiencing the blast effects of the initial round. Firing assessments prove that blast anticipation after firing the initial round causes the firer to concentrate more on blast effects than the target. This can be overcome if Soldiers are given the opportunity to fire more shoulder-launched munitions and at a greater frequency. Soldiers can use simulators that closely replicate the blast effects of firing live munitions to reduce firer anticipation.

Table A-2. Practice day firing tables for M72-series shoulder-launched munitions.

TABLE I—M72 PRACTICE DAY FIRE, STATIONARY TARGETS	
TASK	Engage stationary targets with a M72AS subcaliber training launcher.
CONDITIONS	On a suitable MPRC. Given one M72AS subcaliber training launcher, three rounds of 21-mm ammunition, and three stationary targets at ranges of 150 to 250 meters. One target is engaged while the Soldier is wearing MOPP4 gear.
STANDARD	The Soldier fires three rockets at stationary targets and achieves at least two hits. The Soldier demonstrates correct firing positions, estimates range to the target, and applies the fundamentals of marksmanship.
TABLE II—M72 PRACTICE DAY FIRE, MOVING TARGETS	
TASK	Engage moving targets with a M72AS subcaliber training launcher.
CONDITIONS	On a suitable MPRC. Given one M72AS subcaliber training launcher, two rounds of 21-mm ammunition, and two targets moving at a rate of 8 to 24 km per hour at a range of 100 to 200 meters.
STANDARD	The Soldier fires two rockets at moving targets and achieves at least one hit. The Soldier demonstrates correct firing positions, estimates range to the target, and applies the fundamentals of marksmanship.

A-4.   The results are recorded on Tables I and II of DA Form 7678.

PRACTICE NIGHT FIRE

A-5. Soldiers conduct practice night fire using a M72AS subcaliber training launcher.

Notes.	1.	During the execution of this training, range safety personnel should load the subcaliber training launcher and perform any necessary maintenance.
	2.	Practice night fire consists of hands-on installation of NVDs and firing. Instructors will prepare all shoulder-launched munitions for conducting night fire.

A-6. Table A-3 shows the distribution of rounds.

Table A-3. Distribution of rounds for practice night fire.

ROUND	TYPE OF TARGET	RANGE (METERS)	FIRING POSITION
Practice Stationary/Moving (M72AS Subcaliber Training Launcher)			
1	Stationary	200 to 250	Standing
2	Moving	150 to 200	Kneeling

A-7. Soldiers fire two rounds using a M72AS subcaliber training launcher: one round at a stationary target at a range of 200 to 250 meters and one round at a moving target at a range of 150 to 200 meters. The purpose of this firing is to determine the firer's ability to estimate range to the target during limited visibility conditions, demonstrate correct firing positions, apply the fundamentals of marksmanship, and achieve accuracy. Table A-4 shows the task, conditions, and standards for this training.

Table A-4. Practice night firing table for M72-series shoulder-launched munitions.

TABLE III—M72 PRACTICE NIGHT FIRING	
TASK	Engage stationary targets with a M72AS subcaliber training launcher.
CONDITIONS	On a suitable MPRC. Given one M72AS subcaliber training launcher; two rounds of 21-mm ammunition; two stationary targets at ranges of 125 to 220 meters; one AN/PAS-13 or AN/PVS-4 NVS; NVS mounting bracket; and an AN/PEQ-15, an AN/PAC-4 aiming light, or illumination provided by indirect fire.
STANDARD	The Soldier fires two rockets at stationary and moving targets and achieves at least one hit. The Soldier demonstrates correct firing positions, estimates range to the target, and applies the fundamentals of marksmanship.

Notes.	1.	Before you can use a NVS, you must install the AN/PVS-4 mounting kit. Both sights can be used with this kit.
	2.	Before you can use the AN/PVS-4, you must install a M72A1 sight reticle on it and align it to an expended M72 launcher.
	3.	Before conducting Table III, all NVDs must be aligned to the launcher.

A-8. The results are recorded on Table III of DA Form 7678.

This page intentionally left blank.

Appendix B

TRAINING AIDS, DEVICES, SIMULATORS, AND SIMULATIONS

TADSS eliminate the training gap in the generating force and operational Army by providing a means to conduct hands-on training tasks, marksmanship fundamentals, and target engagements before Soldiers attempt to fire the actual munition.

SECTION I. TRAINING LAUNCHERS

B-1. Shoulder-launched munition training launchers are classified as FHTs or FETs. FHTs are manufactured launchers made to look, feel, and operate the same as a live round. FETs are expended munitions that may or may not feel and operate the same as a live round. Soldiers can use training launchers to practice actions without the added stress of live munitions.

Notes. 1. Shoulder-launched munition training launchers are accountable items; they are tracked by serial number.

2. Refer to DA PAM 385-64, chapter 13, for more information about the approved markings of training launchers.

FIELD-EXPEDIENT TRAINER

B-2. Expended (fired) shoulder-launched munitions (Table B-1) may be converted into FETs as an initial assignment of training rounds and as replacements for damaged rounds. Weight can be added to replicate the actual munition, if needed; however, this type of training launcher may not contain fully functional firer controls. An FET made using an expended launcher should be used for initial shoulder-launched munitions training to help Soldiers learn how to—

- Inspect the launcher for serviceability.
- Prepare the launcher for firing.
- Demonstrate the correct firing position.
- Obtain correct sight picture.
- Perform misfire procedures.
- Return the launcher to a carrying configuration.

Notes. 1. Converted shoulder-launched munitions ARE NOT available through the supply system and are only obtained by direct conversion from expended launchers. Units can submit a request through command channels to keep fired shoulder-launched munitions as training aids. The local training support center (TSC) can provide FETs.

2. DOD Regulation 5100.76-M requires that FETs made from expended launchers to be carefully controlled. Conversion is authorized provided that the unit/organization maintains accountability of the item. Damaged FETs should be disposed of in accordance with unit and post regulations and SOPs. Markings/labels should be used to distinguish training devices from service munitions.

3. The fire controls of expended M141 BDMs are not functional.

Table B-1. Types of field-expedient trainers.

TYPE OF FIELD EXPEDIENT TRAINER	DETAILS
M136 AT4	M136 AT4 FETs are marked with a 1-inch gold band between the front and rear sights, and with the word "INERT" in 1-inch letters on the side of the launcher.
M136A1 AT4CS	M136A1 AT4CS FETs are marked with a 1-inch gold band between the front and rear sights, and with the word "INERT" in 1-inch letters on the side of the launcher.
M72-Series Shoulder-Launched Munitions	M72-series FETs are marked with a 1-inch gold band and have a training label below the munition nomenclature stenciled on the launcher.
M141 BDM	M141 BDM FETs are marked with a 1-inch gold band, and with the word "INERT" in 1-inch letters on the side of the launcher.

FIELD HANDLING TRAINER

B-3. FHTs (Table B-2) come off of the assembly line packaged the same as live rounds, but with all the markings of a training round. An inert ballast is added to replicate the rocket's weight and center of gravity.

Note. The M141 BDM firing mechanism is electrically controlled. A rechargeable firing mechanism that enables the Soldier to recock the launcher must be installed at the factory.

B-4. Fully operational FHTs should be used for basic and advanced shoulder-launched munitions training to help Soldiers learn how to—

- Arm, aim, and fire the launcher during the day.
- Install NVDs and perform sight alignment.
- Arm, aim, and fire the launcher during limited visibility.
- Apply target engagement techniques for stationary and moving targets.

Table B-2. Types of field handling trainers.

TYPE OF FIELD HANDLING TRAINER	DETAILS
M141 BDM	M141 BDM FHTs have gold bands to identify them as trainers. The firing mechanism safety button and trigger button are functional, permitting firers to practice firing. These FHTs can be reset after functioning by using a recocking pin stored under the NVD mount protective cover. **Note.** The recocking mechanism is the only function that differs from the actual round.

SECTION II. SUBCALIBER TRAINING LAUNCHERS

B-5. The subcaliber training launcher is a reusable system that Soldiers can fire from the standing, sitting, kneeling, and prone positions.

WARNING

Subcaliber training launchers look, feel, and behave like actual munitions. For this reason, they should be treated like actual munitions.

Notes. 1. To extend the service life of shoulder-launched munition subcaliber training launchers, they must not be used during classroom instruction, practical exercise drills, or field exercises.

2. Shoulder-launched munition subcaliber training launchers must be serviced and maintained in accordance with their individual TMs. Each subcaliber training launcher kit comes with its own cleaning supplies and TM. Cleaning supplies are in the Army supply system.

3. Shoulder-launched munition subcaliber training launchers are accountable items; they are tracked by serial number.

With the exception of the M136A1 AT4CS and the M72A4/5, each shoulder-launched munition has its own specially designed subcaliber training launcher:

- The M141 BDM currently uses the BDM subcaliber training launcher.
- The M136 AT4 uses the M287 subcaliber training launcher.
- The M72A6/A7 uses the M72AS subcaliber training launcher.

Note. All training with the M190 subcaliber training launcher has been discontinued. Units requesting M72-series munitions can conduct practice live fires with the M72AS subcaliber training launcher. See Appendix A for M72-series shoulder-launched munitions instructional live-fire training tasks.

BUNKER DEFEAT MUNITION SUBCALIBER TRAINING LAUNCHER

B-6. The BDM subcaliber training launcher (Figure B-1) is a specially constructed M141 BDM launcher fitted with a reusable/reloadable, 21-mm subcaliber barrel insert assembly. The technical specifications for this training launcher are listed in Table B-3.

Note. The BDM subcaliber training launcher firing mechanism is electrically controlled. A rechargeable firing mechanism that enables the Soldier to recock the launcher must be installed at the factory.

IDENTIFICATION

B-7. Unlike the live round and the FHT, the BDM subcaliber training launcher has no color-coded band between the front and rear sights.

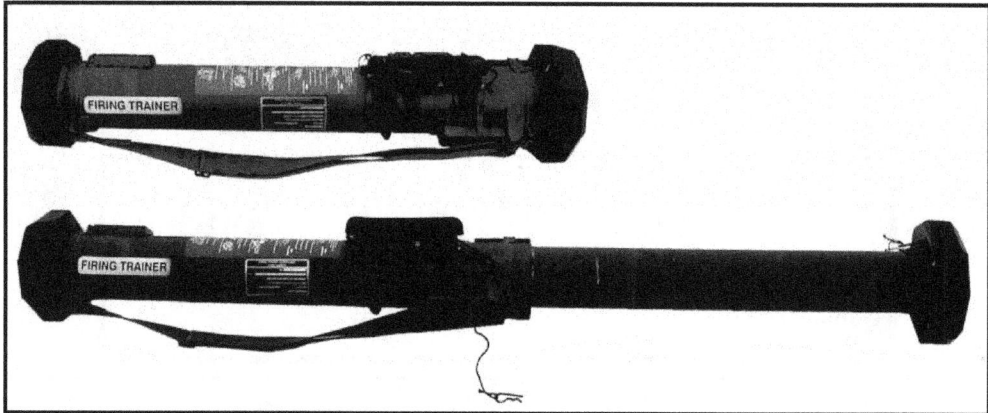

Figure B-1. Bunker defeat munition subcaliber training launcher.

Table B-3. Technical data for the bunker defeat munition subcaliber training launcher.

Length	Closed	32.0 inches (813 mm)
	Extended	54.7 inches (1,389 mm)
Weight	16 pounds (7.26 kg)	
Action	Electrical	
Sights	M136-series shoulder-launched munition front and rear sights	
Operating temperature	-40 to 140 degrees Fahrenheit (-40 to 60 degrees Celsius)	
Muzzle velocity	210 meters per second (690 feet per second)	
Caliber	21-mm	

COMPONENTS

B-8. The training launcher consists of a launcher, a 21-mm subcaliber barrel insert assembly, and a primer block (Figure B-2).

AMMUNITION

B-9. The BDM subcaliber training launcher is designed to accept a HA21 21-mm training rocket (Figure B-3). The velocity and trajectory of this ammunition match those of shoulder-launched munitions. The HA21 21-mm training rocket produces more noise and blast overpressure than the M939 9-mm training practice-tracer (TP-T) cartridge. The HA21 training rocket also has a tracer element that the firer can see out to 250 meters; this enables the firer to compare the impact of the training rocket with the sight picture.

Note. The HA21 training rocket may be fired at stationary or moving targets. It can be fired on MPRCs.

Figure B-2. Bunker defeat munition subcaliber training launcher primer block.

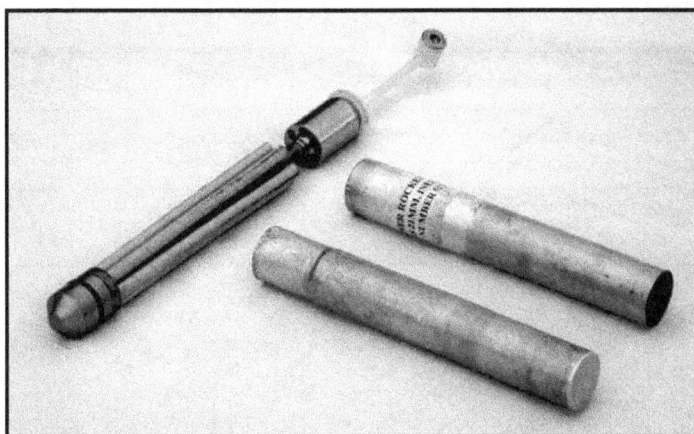

Figure B-3. HA21 21-mm training rocket with storage case.

FUNCTION CHECK

Note. The BDM subcaliber training launcher comes packaged in the carrying configuration.

B-10. Before the BDM subcaliber training launcher is fired, a function check must be performed to ensure the trigger and safety mechanisms are operating properly. To perform a function check, ensure—

- The launcher has no apparent damage. Check carefully for cracks or breaks to the firing mechanism and the front and rear sight covers.
- The firing mechanism and sight cover open and close.
- The recocking and safety pins are present.
- The launcher extends and locks in the extended position, and the front and rear sights lock in the upright position.
- The safety button and firing pin are in proper working condition. Check by cocking and firing the unloaded training launcher.

CAUTION

Do not dry-fire without an expended rocket primer block. It may damage the firing mechanism.

- The primer block cover/housing cavity has no apparent damage.
- The bore gauge will pass through the barrel assembly.

WARNINGS

Load live ammunition when on the firing line only.

Never fire the BDM subcaliber training launcher at hard targets less than 100 meters from the firing line.

Remain clear of the front of the launcher, which must be pointed downrange at all times.

B-11. Soldiers should return damaged/unserviceable BDM subcaliber training launchers to the issue point.

Note. If the 21-mm subcaliber barrel insert assembly is damaged, field-level maintenance can replace its complete firing mechanism.

COCKING

Notes. 1. The BDM subcaliber training launcher requires the firing mechanism to be cocked prior to use and after each firing sequence. This requires the launcher to be collapsed and re-extended after each firing.

2. The BDM subcaliber training launcher comes packaged in the carrying configuration. The initial cocking (performed before placing the launcher in the ready-to-fire configuration) and the recocking of the M141 subcaliber training launcher firing mechanism will be performed by assistant firers or range personnel.

B-12. See Table B-4 for more information about the procedures used to cock the BDM subcaliber training launcher.

Table B-4. Procedures used to cock the bunker defeat munition subcaliber training launcher.

FIRER	ASSISTANT
WARNING Keep the training launcher pointed downrange at all times.	
(1) Hold the launcher on your shoulder (muzzle end pointed downrange, toward the target).	

**Table B-4. Procedures used to cock
the bunker defeat munition subcaliber training launcher (continued).**

FIRER	ASSISTANT
	(1) Remove the recocking pin from under the NVD mounting bracket cover.
	(2) Open the firing mechanism cover.
	(3) Insert the recocking pin into the recocking pin hole on the firing mechanism.
	(4) Close the firing mechanism cover until it touches the recocking pin.
	(5) Push the recocking pin forward until the firing mechanism snaps into the cocked position.
	(6) Remove the recocking pin.
	(7) Close the firing mechanism cover over the trigger button (SAFE position).
	(8) Store the recocking pin under the NVD mounting bracket cover.

PREPARE THE LAUNCHER FOR FIRING

B-13. See Table B-5 for more information about the procedures used to prepare the BDM subcaliber training launcher for firing.

**Table B-5. Procedures used to prepare
the bunker defeat munition subcaliber training launcher for firing.**

FIRER	ASSISTANT
WARNING **Keep the training launcher pointed downrange at all times.**	
(1) Cradle the munition (muzzle end pointed downrange, toward the target).	
(2) While supporting the launcher with your right arm, place the launcher under your left arm.	
(3) Hold the launcher away from the body.	
(4) While keeping the muzzle of the launcher pointed downrange, face to the rear by pivoting your body 90 degrees to the right.	
(5) With your right hand, pull and release the transport safety pin.	
(6) Depress the tube release button with your left thumb.	
(7) Grasp the rear tube (inner tube) just in front of the rear bumper with your right hand, and extend the inner tube rearward until it stops.	
Note. A yellow band is visible at the inner tube front end when the tube is fully extended.	
(8) Release the tube release button.	
(9) Rotate the inner tube clockwise (in the direction of the arrow) until it locks.	
(10) Verify that the inner tube is locked by attempting to rotate the inner tube counterclockwise (in the opposite direction of the arrow).	
Note. If the tubes are not locked, the munition will not arm.	
(11) Inspect the inner tube for cracks, dents, or punctures. If any are present, return the launcher to its carrying configuration, tag it, and return it to the TSC.	
(12) Press the shoulder stop lock/release button, and pull the shoulder stop out.	
(13) Grip the forward end of the launcher with your left hand and the rear end of the launcher with your right hand.	
(14) Raise the launcher out and away from your body.	
(15) While keeping the launcher pointed at the target, pivot your body 180 degrees to face the target.	
(16) Place the launcher on your right shoulder.	

**Table B-5. Procedures used to prepare
the bunker defeat munition subcaliber training launcher for firing (continued).**

FIRER	ASSISTANT
(17) Reach forward with your left hand, and grasp the front sight cover. Press down, and slide it rearward. (18) With your left hand, grasp the rear sight cover. Press down, and slide it forward. (19) Wrap the sling strap around your left bicep. Cup the bottom of the launcher with your left hand, and slide it back toward your body to tighten the sling. --- ***Note.*** When firing the M141 BDM, the weapon sling should be used to increase firer control, as is done with a conventional rifle; however, DO NOT wrap the sling around your left arm as one would with a rifle.	

LOADING

B-14. See Table B-6 for more information about the procedures used to load the BDM subcaliber training launcher.

Table B-6. Procedures used to load the bunker defeat munition subcaliber training launcher.

FIRER	ASSISTANT
WARNING Keep the training launcher pointed downrange at all times. (1) Hold the launcher on your shoulder (muzzle end pointed downrange, toward the target).	(1) Place the safety pin into the hole in training launcher's rear tube.

Table B-6. Procedures used to load
the bunker defeat munition subcaliber training launcher (continued).

FIRER	ASSISTANT
	(2) Grasp the training rocket's storage case, and separate the two halves.

WARNINGS

Wear gloves when handling HA21 21-mm training rockets.

To prevent electrostatic discharge, a bare HA21 21-mm training rocket should never be handed from one person to another.

The HA21 21-mm training rocket should remain in the aluminum storage tube until just prior to loading in the training launcher.

Do not grip the propellant portion of the HA21 21-mm training rocket when removing the training rocket from the carrying case or when loading into the BDM subcaliber training launcher. The training rocket propellant may be damaged, causing a misfire.

(3) Carefully remove the training rocket from the case. Inspect the training rocket for—
- Broken or missing propellant sticks.
- Broken igniter or transfer line.
- Damaged or missing O-ring.
- Dirt and debris.

**Table B-6. Procedures used to load
the bunker defeat munition subcaliber training launcher (continued).**

FIRER	ASSISTANT
	  (4) Rotate the primer block cover open to expose the primer housing cavity. (5) Remove the dust cover from the primer housing cavity using a flat-tipped screw driver.

Table B-6. Procedures used to load
the bunker defeat munition subcaliber training launcher (continued).

FIRER	ASSISTANT
	  WARNING The assistant MUST keep his body out of the backblast area when inserting the HA21 21-mm training rocket primer block in the housing of the BDM subcaliber training launcher. (6) While holding the training rocket by the nozzle end, line up the primer block on the training rocket with the BDM subcaliber training launcher primer housing cavity. (7) Insert the training rocket into the training launcher barrel until the flange of the training rocket is flush against the subcaliber training launcher. *Note.* Ensure that the training rocket is fully inserted in the launch tube before placing the primer in the primer housing.

**Table B-6. Procedures used to load
the bunker defeat munition subcaliber training launcher (continued).**

FIRER	ASSISTANT
	(8) Place the primer block in the housing. (9) Close the primer block cover over the training rocket primer block. (10) While standing to the firing side of the launcher, remove the safety pin from the training launcher rear tube. Let the firer know the training launcher is ready to be fired.

ARMING/FIRING

B-15. See Table B-7 for more information about the procedures used to arm and fire the BDM subcaliber training launcher.

RELOADING

B-16. See Table B-8 for more information about the procedures used to reload the BDM training launcher.

Table B-7. Procedures used to arm/fire
the bunker defeat munition subcaliber training launcher.

FIRER	ASSISTANT
WARNING Keep the training launcher pointed downrange at all times. Check the backblast area before firing the munition. When firing at targets less than 100 meters away, fire from behind safe cover (i.e., sandbag wall, concrete wall) to prevent injury from flying debris. (1) Hold the launcher on your shoulder (muzzle end pointed downrange, toward the target). (2) Ensure the backblast area is clear of personnel. (3) Grasp the firing mechanism cover with your right hand, and rotate the cover all the way forward until the cover is flush with the outer tube. *Notes.* 1. If the firing mechanism cover is not flush with the launch tube, the munition will not arm. 2. The word ARMED can be seen in red letters when the cover is opened. (4) Adjust the rear sight to the correct range, using the following: *Note.* When opening the rear sight cover, the range is preset at the 150-meter battlesight range setting. • To adjust the rear sight range setting to more than 150 meters, turn the range knob clockwise (toward the muzzle). • To decrease the range, turn the range knob counterclockwise (toward the firer). *Note.* There is an audible clicking sound at each 50-meter increment; this sound aids you during limited visibility. (5) Place the fingertips of your right hand on the safety button (located on top of the firing mechanism), and press down. Then, place your right thumb on the red trigger button. (6) Pull the shoulder stop against the shoulder. (7) Aim the launcher. *Note.* The rear sight should be no less than 2 1/2 inches and no more than 3 inches from your eyes. (8) Press the trigger button forward with the thumb of your right hand, and hold until the subcaliber training launcher fires.	

Table B-8. Procedures used to reload the bunker defeat munition subcaliber training launcher.

FIRER	ASSISTANT
WARNING Keep the training launcher pointed downrange at all times. (1) Hold the launcher on your shoulder (muzzle end pointed downrange, toward the target).	(1) Open the firing mechanism cover. (2) Insert the recocking pin into the recocking pin hole on the firing mechanism. (3) Close the firing mechanism cover until it touches the recocking pin. (4) Push the recocking pin forward until the firing mechanism snaps into the cocked position. (5) Remove the recocking pin. (6) Close the firing mechanism cover over the trigger button (SAFE position). (7) Store the recocking pin under the NVD mounting bracket cover.
(2) Keeping the launcher's muzzle pointed toward the target area, remove the training launcher from your shoulder, and place the launcher under your left arm as you pivot your body 180 degrees to face to the rear. (3) Depress the tube release button with your left thumb. (4) Rotate the inner tube counterclockwise (opposite to the arrow). (5) Release the tube release button. (6) With your right hand, insert the inner tube into the outer tube until about 6 inches of the inner tube is still exposed. (7) Grasp the rear tube (inner tube) just in front of the rear bumper with your right hand, and extend the inner tube rearward until it stops. (8) Rotate the inner tube clockwise (in the direction of the arrow) until it locks. (9) Verify that the inner tube is locked by attempting to rotate the inner tube counterclockwise (opposite to the arrow). Grip the forward end of the launcher with your left hand and the rear end of the launcher with your right hand. (10) While keeping the launcher pointed at the target, raise the launcher out and away from your body, pivot your body 180 degrees to the left, and place the launcher on your right shoulder, facing the target area.	

**Table B-8. Procedures used to reload
the bunker defeat munition subcaliber training launcher (continued).**

FIRER	ASSISTANT
(11) Wrap the sling strap around your left bicep. Cup the bottom of the launcher with your left hand, and slide it back toward your body to tighten the sling. *Note.* When firing the M141 BDM, the weapon sling should be used to increase firer control, as is done with a conventional rifle; however, DO NOT wrap the sling around your left arm as one would with a rifle.	(8) Rotate the primer block cover open to expose the primer housing cavity. (9) Remove the expended primer cap from the primer housing cavity. **CAUTION** Use only approved solvent and/or lubricant to clean the training launcher before firing and after each three rounds fired. (10) Inspect the primer block housing cavity for debris and carbon build-up. Clean with a damp rag or cleaning kit, as necessary. *Note.* If required, the firer cradles the launcher enabling the assistant to clean the barrel after each three rounds fired, while the firer keeps the muzzle end pointed downrange, toward the target. (11) Place the safety pin in the training launcher rear tube.

Table B-8. Procedures used to reload
the bunker defeat munition subcaliber training launcher (continued).

FIRER	ASSISTANT
	WARNINGS Wear gloves when handling HA21 21-mm training rockets. To prevent electrostatic discharge, a bare HA21 21-mm training rocket should never be handed from one person to another. The HA21 21-mm training rocket should remain in the aluminum storage tube until just prior to loading in the training launcher. Do not grip the propellant portion of the HA21 21-mm training rocket when removing the training rocket from the carrying case or when loading into the BDM subcaliber training launcher. The training rocket propellant may be damaged, causing a misfire.
	(12) Grasp the training rocket's storage case, and separate the two halves. Carefully remove the training rocket from the case.
	(13) Inspect the training rocket for— • Broken or missing propellant sticks. • Broken igniter or transfer line. • Damaged or missing O-ring. • Dirt and debris.
(12) Prepare the launcher for firing.	(14) Load the training launcher.
	Note. See the Loading section of this appendix for more information about loading the training launcher.
(13) Check the backblast area.	
(14) Grasp the firing mechanism cover with your right hand, and rotate the cover all the way forward until the cover is flush with the outer tube.	
Notes. 1. If the firing mechanism cover is not flush with the launch tube, the munition will not arm. 2. The word ARMED can be seen in red letters when the cover is opened.	
(15) Ensure the rear sight is set to the correct range.	
(16) Place the fingertips of your right hand on the safety button (located on top of the firing mechanism), and press down. Then, place your right thumb on the red trigger button.	
(17) Pull the shoulder stop against the shoulder.	
(18) Aim the launcher.	
(19) Press the trigger button forward with the thumb of your right hand, and hold until the subcaliber training launcher fires.	

UNLOADING (PREPARED BUT NOT FIRED)

B-17. See Table B-9 for more information about the procedures used to unload an unexpended HA21 training rocket from the BDM subcaliber training launcher. These procedures are used to remove the training rocket if the subcaliber training launcher was prepared for firing, but no firing actions were initiated.

Table B-9. Procedures used to reload the bunker defeat munition subcaliber training launcher.

FIRER	ASSISTANT
WARNING Keep the training launcher pointed downrange at all times.	**WARNING** Keep your body out of the backblast area when inserting the BDM subcaliber training launcher safety pin.
(1) Close the firing mechanism cover (SAFE position), and wait for further instructions from the assistant.	
	(1) Place the safety pin in the training launcher rear tube.
	(2) Open the primer block cover to expose the training rocket primer block.
	(3) Carefully grasp the flash tube and pull rearward, removing the primer block and training rocket from the subcaliber training launcher.
	(4) Return the training rocket to its aluminum case, and return the training rocket to the ASP at the end of the training day.
	(5) Instruct the firer to restore the training launcher to the carrying configuration.

PERFORMING MISFIRE PROCEDURES

Note. BDM subcaliber training launcher misfires must be treated as if firing a live munition. Unlike live munitions, misfire procedures are performed by both the firer and the assistant.

B-18. See Table B-10 for more information about the procedures used to address a misfire.

**Table B-10. Procedures used to address a misfire on
the bunker defeat munition subcaliber training launcher.**

FIRER	ASSISTANT
WARNING If a misfire occurs, wait 90 seconds. Do not remove the training launcher from the shoulder, and keep the training launcher pointed downrange at all times.	**WARNING** Keep your body out of the backblast area when inserting the BDM subcaliber training launcher safety pin.

<table>
<tr><td>

(1) Immediately announce misfire.

(2) Keep the launcher pointed toward the target.

(3) Release the trigger button and safety button.

(4) Resqueeze the safety button firmly. Hold and aim. Press the trigger button.

(5) If the training launcher fails to fire, say "MISFIRE." Close the firing mechanism cover (SAFE position).

(6) Check the backblast area.

(7) Open the firing mechanism cover, flush with the tube. Squeeze and hold the safety button. Aim. Press the trigger button.

(8) If the training launcher fails to fire again, say "MISFIRE." Maintain the firing position for 90 seconds.

(9) After 90 seconds have passed, close the firing mechanism cover (SAFE position), and wait for further instructions from the assistant.

</td><td>

(1) Replace the safety pin in the training launcher rear tube.

Note. If the safety pin cannot be replaced in the hole in the training launcher rear tube, the assistant will notify the range NCOIC.

(2) Open the primer housing cavity.

</td></tr>
</table>

**Table B-10. Procedures used to address a misfire on
the bunker defeat munition subcaliber training launcher (continued).**

FIRER	ASSISTANT
	(3) Remove the M21 training rocket from the training launcher by grasping the flange.
	Note. To remove the training rocket from the subcaliber training launcher, refer to Unloading portion of this section.
	(4) Inspect the primer cap for evidence of contact with the firing pin. • If the primer cap shows evidence of contact with the firing pin, reinsert the training rocket into its metallic case and notify safety personnel. • If the primer cap shows no evidence of contact with the firing pin, perform loading operations and firing procedures.

M287 SUBCALIBER TRAINING LAUNCHER

B-19. The M287 subcaliber training launcher (Figure B-4) is a specially constructed M136 AT4 launcher fitted with a reusable/reloadable 9-mm subcaliber barrel insert assembly. The technical specifications for this training launcher are listed in Table B-11.

Notes. 1. The M287 subcaliber training launcher is not made from an expended launcher, so it has its own national stock number (NSN). This subcaliber training launcher is available through the Army supply system.

2. See TM 9-1055-886-12&P for more information.

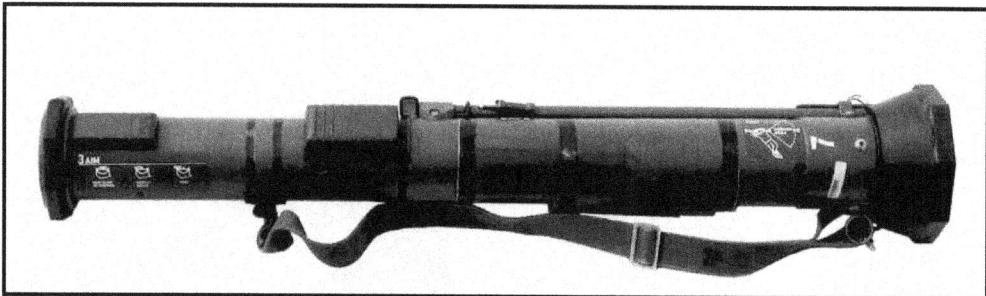

Figure B-4. M287 subcaliber training launcher.

Table B-11. Technical data for the M287 subcaliber training launcher.

Length	40 inches (1,020 mm)
Weight	15 pounds (7 kg)
Action	Mechanical
Sights	M136-series shoulder-launched munition front and rear sights
Operating temperature	0 to 100 degrees Fahrenheit (10 to 27 degrees Celsius)
Muzzle velocity	300 meters per second (984 feet per second)
Caliber	9-mm

IDENTIFICATION

B-20. Unlike the live round and the FHT, the M287 subcaliber training launcher has no color-coded band between the front and rear sights.

COMPONENTS

B-21. The training launcher consists of a launcher, a 9-mm subcaliber barrel insert assembly, a breech assembly, and a bolt (Figure B-5). The bolt is easily removed to load M939 9-mm TP-T cartridges and to inspect the barrel for obstructions.

Note. The bolt is the only subcaliber part removed by the operator.

Figure B-5. M287 subcaliber training launcher bolt.

AMMUNITION

B-22. The M287 training launcher is designed to accept a special rifle barrel that fires the M939 9-mm TP-T cartridge (Figure B-6). The velocity and trajectory of this ammunition match those of the M136 AT4's HEAT cartridge, but the M287 training launcher produces less noise, has minimal backblast, and minimal overpressure. The M939 9-mm TP-T cartridge has a lighter powder charge than a standard 9-mm bullet. The lighter charge enables the cartridge to closely replicate the trajectory of the M136 AT4 tactical round at ranges out to 700 meters. The M939 TP-T cartridge also has a tracer element to enable the firer to compare the impact of the cartridge with the sight picture. The firer can see the tracer out to 450 meters.

WARNING

The M939 9-mm TP-T cartridge's red tip distinguishes it from standard 9-mm ammunition, which should never be fired from the M287 subcaliber training launcher.

Note. The M939 9-mm TP-T cartridge may be fired at stationary or moving targets. It can be fired on MPRCs and at tanks; however, before it can be fired at a tank occupied by personnel, the parts of the tank that could suffer damage must be shielded. The local TSC can provide specifications for modifying tanks to be used as targets for the M287 subcaliber training launcher.

Figure B-6. M939 9-mm training practice-tracer cartridge.

FUNCTION CHECK

B-23. Before the M287 subcaliber training launcher is fired, a function check must be performed to ensure the trigger and safety mechanisms are operating properly. Before performing a function check, ensure—

- The cocking lever is in the SAFE ("S") position.
- The transport safety pin is fully inserted, with the lanyard wrapped clockwise around the launcher.
- The bolt is removed from the breech.

B-24. Function check procedures are shown in Table B-12.

Note. If the M287 subcaliber training launcher is damaged, field-level maintenance can replace its complete firing mechanism.

Table B-12. Function check, M287 subcaliber training launcher.

STEP	OPERATOR ACTIONS	FUNCTION CHECK	CORRECTIVE MEASURES
1	Try to cock the subcaliber training launcher with the transport safety pin installed.	It should not cock.	If it cocks, turn the training launcher in for repair.
2	Remove the transport safety pin, and depress the forward safety (red).	The forward safety should spring back when released.	If it does not spring back, turn the training launcher in for repair.
3	Cock the firing mechanism, pressing only the red trigger button.	The firing rod should not protrude through the rear of the firing tube assembly.	If the training launcher fires, turn it in for repair.
4	Recock the firing mechanism. Fully depress, and continue to hold down the forward safety. Press the red trigger button.	The firing mechanism should function. The firing rod should protrude through the rear of firing tube assembly.	If the training launcher fails to fire, turn it in for repair.
5	Return the cocking lever to the SAFE position and reinstall the transport safety pin.		

WARNINGS

Load live ammunition when on the firing line only.

Never fire the M287 subcaliber training launcher at hard targets less than 125 meters from the firing line.

Remain clear of the front of the launcher, which must be pointed downrange at all times.

LOADING

Note. The M287 subcaliber training launcher comes packaged in the carrying configuration.

B-25. See Table B-13 for more information about the procedures used to load the M287 subcaliber training launcher.

Table B-13. Procedures used to load the M287 subcaliber training launcher.

FIRER	ASSISTANT
(1) Cradle the launcher. (2) While supporting the launcher with your left hand, pull and release the transport safety pin with your right hand. (3) Unsnap, unfold, and hold the shoulder stop with your right hand. (4) Grip the base of the sling on the front of the launcher with your left hand and the shoulder stop with your right hand. (5) Raise the munition out and away from your body. (6) While keeping the munition pointed at the target, pivot your body 90 degrees to face the target. (7) Place the munition on your right shoulder. *Note.* You can use the carrying strap to steady the munition. (8) Reach forward with your right hand, and grasp the front sight cover. Press down, and slide it rearward. (9) With your right hand, grasp the rear sight cover. Press down, and slide it forward. (10) Pull back on the sling with your left hand to seat the shoulder stop firmly against your shoulder, and hold. (11) Hold the launcher on your shoulder (muzzle end pointed downrange, toward the target). *Note.* Place your firing hand by your side while the assistant is loading and unloading the launcher.	

Table B-13. Procedures used to load the M287 subcaliber training launcher (continued).

FIRER	ASSISTANT
	(1) Inspect the 9-mm TP-T cartridge primer to ensure it is not dented. (2) Remove the bolt by turning it counterclockwise, past the SAFE position ("S"). (3) Pull the bolt from the breech. (4) Look through the barrel from the rear to verify that it contains no obstructions. (5) Slide the primer end of the 9-mm TP-T cartridge into the slotted groove of the bolt.

Table B-13. Procedures used to load the M287 subcaliber training launcher (continued).

FIRER	ASSISTANT
	(6) Insert the bolt into the breech. (7) Press and turn the bolt clockwise to the SAFE position ("S"). *Note.* The cocking lever must be in the SAFE position for the bolt to turn to the FIRE position. (8) Turn the bolt clockwise to the FIRE position ("F").

Table B-13. Procedures used to load the M287 subcaliber training launcher (continued).

FIRER	ASSISTANT
(12) Ensure the backblast area is clear of personnel. (13) Unfold the cocking lever with your right hand. Place your thumb under it and, with the support of your fingers in front of the firing mechanism, push it forward, rotate it downward and to the right, and let it slide backward. (14) Adjust the rear sight to the correct range, using the following procedures: *Note.* When opening the rear sight cover, the range is preset at the 150-meter battlesight range setting. • To adjust the rear sight range setting to more than 150 meters, turn the range knob clockwise (toward the muzzle). • To decrease the range, turn the range knob counterclockwise (toward the firer). *Note.* There an audible clicking sound at each 50-meter increment; this sound aids you during limited visibility. (15) Place the first two fingers of your right hand on the red safety release catch, and extend the thumb. While keeping the thumb extended, press the red safety release catch down, and hold. (16) Ensure the shoulder stop is firmly against the shoulder, and hold. (17) Aim the launcher. *Note.* The rear sight should be no less than 2 1/2 inches and no more than 3 inches from your eyes. (18) Press the red trigger button forward with the thumb of your right hand and hold until the subcaliber training launcher fires. 	

RELOADING

> *Note.* The assistant firer performs these actions, while the firer holds the launcher on his shoulder (muzzle end pointed downrange, toward the target).

B-26. See Table B-14 for more information about the procedures used to reload the M287 subcaliber training launcher.

Table B-14. Procedures used to reload the M287 subcaliber training launcher.

FIRER	ASSISTANT
WARNING Keep the training launcher pointed downrange at all times.	**DANGER** NEVER TOUCH THE TRIGGER WHILE RELOADING.
DANGER NEVER TOUCH THE TRIGGER WHILE RELOADING.	
(1) Hold the launcher on your shoulder (muzzle end pointed downrange, toward the target).	
(2) Place the cocking lever in the SAFE position ("S"), and place your firing hand by your side.	
	(1) Reload the subcaliber training launcher by following Steps 1 through 8 for loading (as indicated for the assistant).
(3) Prepare the launcher for firing by following Steps 12 through 18 for loading (as indicated for the firer).	

UNLOADING (PREPARED BUT NOT FIRED)

WARNING

Keep the training launcher pointed downrange at all times.

B-27. See Table B-15 for more information about the procedures used to unload the M287 subcaliber training launcher (Figure B-7). These procedures are used to remove the training rocket if the subcaliber training launcher was prepared for firing, but no firing actions were initiated.

Figure B-7. Unloading the M287 subcaliber training launcher.

Table B-15. Procedures used to unload the M287 subcaliber training launcher.

FIRER	ASSISTANT
WARNING **Keep the training launcher pointed downrange at all times.**	
(1) Cradle the launcher or place the launcher on your shoulder (muzzle end pointed downrange, toward the target).	
(2) Place your firing hand by your side while the assistant is loading and unloading the launcher.	
	(1) Remove the bolt by turning it counterclockwise, past the SAFE position ("S").
	(2) Pull the bolt from the breech.
	(3) If more rounds are to be fired, remove the expended 9-mm TP-T cartridge, and reload the bolt by following Steps 1 through 8 for loading.
	(4) Reload the subcaliber training launcher by following Steps 1 through 8 for loading (as indicated for the assistant).
(3) Prepare the launcher for firing by following Steps 12 through 18 for loading (as indicated for the firer).	

PERFORMING MISFIRE PROCEDURES

Note. M287 subcaliber training launcher misfires must be treated as if firing a live munition. Unlike live munitions, misfire procedures are performed by both the firer and the assistant.

B-28. See Table B-16 for more information about the procedures used to address a misfire.

Table B-16. Procedures used to address a misfire on the M287 subcaliber training launcher.

FIRER	ASSISTANT
WARNING **Keep the training launcher pointed downrange at all times.** (1) Hold the launcher on your shoulder (muzzle end pointed downrange, toward the target). *Note.* The firer performs these actions while holding the launcher on his shoulder (muzzle end pointed downrange, toward the target). (2) If the munition does not fire, announce "MISFIRE". (3) Release the red trigger button and the red safety release catch. (4) Wait five seconds. Remove your right hand from the firing mechanism, check the backblast area, and cock the munition again. *Note.* Count the seconds by saying "one thousand and one, one thousand and two," and so on. (5) Press down on the red safety release catch firmly, and hold. (6) Aim the munition. (7) Press and hold the red trigger button. (8) If the munition does not fire, announce "MISFIRE". (9) Release the red trigger button and red safety release catch. (10) Maintain the firing position for two minutes, and return the cocking lever to the SAFE (uncocked) position. (11) Place your firing hand by your side while the assistant is loading and unloading the launcher.	

(12) Prepare the launcher for firing by following Steps 12 through 18 for loading (as indicated for the firer).	(1) Ensure the firer placed the cocking lever in the SAFE position ("S") and that the forward safety is in the vertical position.
	
	(2) Insert the transport safety pin, and remove and inspect the 9-mm TP-T cartridge.
	   • If the primer is dented, replace it, and dispose of the old one in accordance with the range safety SOP.
	   • If the primer is not dented, notify field-level maintenance to inspect the bolt firing pin for damage.
	
	(3) Reload the subcaliber training launcher by following Steps 1 through 8 for loading.

M72AS SUBCALIBER TRAINING LAUNCHER

B-29. The M72AS subcaliber training launcher (Figure B-8) is a specially constructed M72A6/A7 shoulder-launched munition launcher fitted with a reusable/reloadable, 21-mm subcaliber barrel insert assembly. The technical specifications for this training launcher are listed in Table B-17.

Notes. 1. Expended M72A1/A2/A3 and M72A6/A7 shoulder-launched munitions are not to be converted to training launchers for the M72AS subcaliber training launcher.

2. This subcaliber training launcher is available through the Army supply system.

Table B-17. Technical data for M72AS subcaliber training launcher.

Length	Closed	31 inches
	Extended	39 inches
Weight	8 pounds	
Trigger load (minimum)	3 pounds	
Sights	Rifle-type rear peep and front post	
Operating temperature	-40 to 140 degrees Fahrenheit (-40 to 60 degrees Celsius)	
Muzzle velocity	221 meters per second	
Caliber	21-mm	

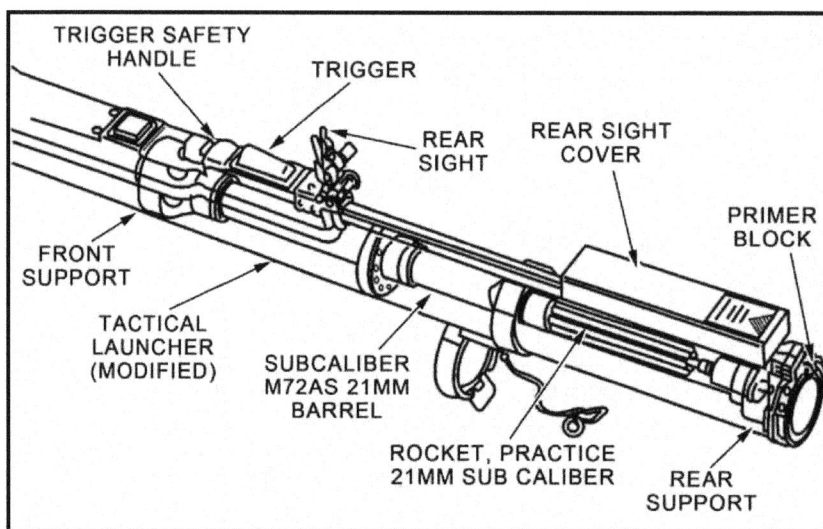

Figure B-8. M72AS subcaliber training launcher.

IDENTIFICATION

B-30. Like the live round and the FHT, the M72AS subcaliber training launcher has a label that identifies it as a training launcher.

COMPONENTS

B-31. The training launcher consists of a launcher, a 21-mm subcaliber barrel insert assembly, and a primer block (Figure B-9).

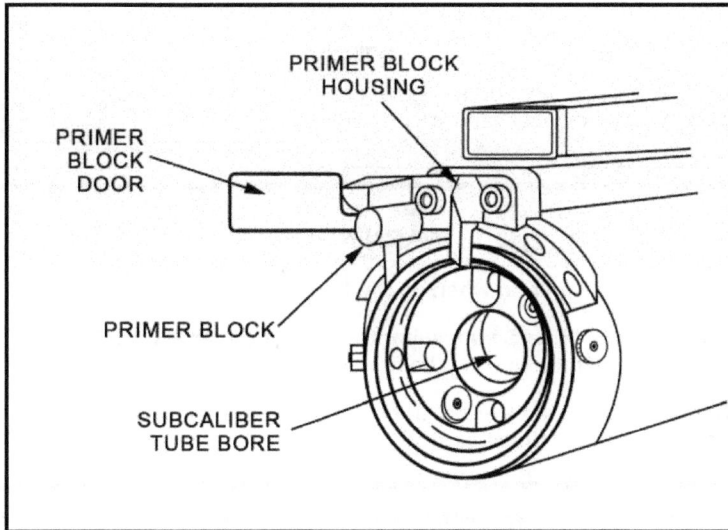

Figure B-9. M72AS 21-mm primer block.

AMMUNITION

B-32. The M72AS training launcher is designed to accept a HA21 21-mm training rocket (Figure B-10). The velocity and trajectory of this ammunition match those of M136- and M72-series shoulder-launched munitions, and the M141 BDM. The HA21 21-mm training rocket produces more noise and blast overpressure than the M939 9-mm cartridge. The HA21 training rocket also has a tracer element that the firer can see out to 250 meters; this enables the firer to compare the impact of the training rocket with the sight picture.

Note. The HA21 training rocket may be fired at stationary or moving targets. It can be fired on MPRCs.

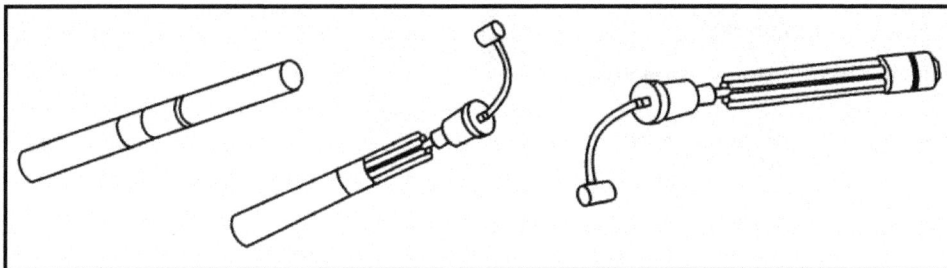

Figure B-10. HA21 21-mm training rocket with storage case.

FUNCTION CHECK

B-33. Before the M72AS subcaliber training launcher is fired, a function check must be performed to ensure the trigger and safety mechanisms are operating properly. To perform a function check, ensure—

- The launcher has no apparent damage. Check carefully for cracks or breaks to the firing mechanism.
- The primer block cover/housing cavity has no apparent damage.
- The launcher extends and locks in the extended position.
- The front and rear sights lock in the upright position and are not damaged.

CAUTION

Do not dry-fire without an expended rocket primer block. It may damage the firing mechanism.

WARNINGS

Load live ammunition when on the firing line only.

Never fire the M72AS subcaliber training launcher at hard targets less than 100 meters from the firing line.

Remain clear of the front of the launcher, which must be pointed downrange at all times.

B-34. Soldiers should return damaged/unserviceable M72AS subcaliber training launchers to the issue point.

Note. If the M72AS subcaliber training launcher is damaged, field-level maintenance can replace its complete firing mechanism.

LOADING

B-35. See Table B-18 for more information about the procedures used to load the M72AS subcaliber training launcher.

Table B-18. Procedures used to load the M72AS subcaliber training launcher.

FIRER	ASSISTANT
WARNING Keep the training launcher pointed downrange at all times. (1) Cradle the launcher, and face the target. (2) Place the launcher on your shoulder. **WARNING** Do not extend the M72AS training launcher to the cocked position until after the 21-mm training rocket is completely installed.	

Table B-18. Procedures used to load the M72AS subcaliber training launcher (continued).

FIRER	ASSISTANT
	(1) Grasp the training rocket metallic case, and separate the two halves. (2) Carefully remove the training rocket from the metallic case. KEEP ROCKET POINTED DOWNRANGE **WARNINGS** To prevent electrostatic discharge, a bare HA21 21-mm training rocket should never be handed from one person to another. The HA21 21-mm training rocket should remain in the aluminum storage tube until just prior to loading in the training launcher. Do not grip the propellant portion of the HA21 21-mm training rocket when removing the training rocket from the carrying case or when loading into the M72AS subcaliber training launcher. The training rocket propellant may be damaged, causing a misfire.

Table B-18. Procedures used to load the M72AS subcaliber training launcher (continued).

FIRER	ASSISTANT
	(3) Inspect the training rocket for— • Broken or missing propellant sticks. • Broken igniter/transfer line. • Damaged or missing O-ring. • Dirt and debris. (4) Swing the primer block cover open to expose the primer housing cavity on the partially collapsed subcaliber training launcher. PRIMER BLOCK COVER PRIMER HOUSING CAVITY **WARNING** **Keep your body out of the backblast area when inserting the HA21 21-mm training rocket primer block in the housing of the M72AS subcaliber training launcher.** (5) Holding the training rocket by the nozzle end, line up the primer block on the training rocket with the subcaliber training launcher primer housing.

Table B-18. Procedures used to load the M72AS subcaliber training launcher (continued).

FIRER	ASSISTANT
	(6) Insert the training rocket into the subcaliber training launcher barrel, until the flange is fully against the subcaliber training launcher.

Note. Ensure that the training rocket is fully inserted in the launch tube before placing the primer in the primer housing.

(7) Close the primer block cover over the training rocket primer block.

ARMING/FIRING

B-36. See Table B-19 for more information about the procedures used to arm and fire the M72AS subcaliber training launcher.

Table B-19. Procedures used to arm/fire the M72AS subcaliber training launcher.

FIRER	ASSISTANT

(1) Hold the training launcher slightly away from the body, and grasp the rear sight cover firmly.
 	
(2) Extend the training launcher by moving the hands firmly in opposite directions.
 	
Notes.   1. Extending the training launcher too slowly can result in failure to cock the training launcher.
 	
       2. Ensure the training launcher snaps into the locked position. Check the tube lock by trying to push the tube together. The training launcher should not collapse if the training launcher is in the locked position.
 	
       3. If two attempts to lock the training launcher fail, close and set the training launcher aside for disposal by authorized ammunition personnel.
 	
(3) Place the training launcher on the firing shoulder.
 	
(4) Check the backblast area.
 	
(5) Arm the training launcher, and wait for further instructions from the assistant.	

	(1) Let the firer know it is ready to be fired.

RELOADING

B-37. See Table B-20 for more information about the procedures used to reload the M72AS subcaliber training launcher.

Table B-20. Procedures used to reload the M72AS subcaliber training launcher.

FIRER	ASSISTANT
WARNING **Keep the training launcher pointed downrange at all times.** (1) With the training launcher on the firing shoulder, return the trigger arming handle to the SAFE position. (2) Remove the training launcher from the shoulder. **CAUTION** Do not damage the front and rear sight when closing the M72AS training launcher. (3) Collapse the inner tube into the outer tube. *Note.* If the trigger safety handle will not return to the SAFE position after firing the training launcher, partially collapse the training launcher by depressing the detent boot, and then depress the trigger bar. An alternate method is to collapse the training launcher, pull forward on the trigger safety handle, and release it.	(1) Open the primer block door. (2) Remove the fired primer block from the primer block housing cavity. **CAUTION** Use only approved solvent and/or lubricant to clean the training launcher before and after each time the launcher is fired. (3) Inspect the primer block housing cavity for debris and/or carbon build-up. Clean with a damp rag or cleaning kit, as necessary. (4) Reload the launcher, as necessary. *Notes.* 1. Keep the training launcher clean between firings. 2. To reload the HA21 training rocket, refer to the Loading portion of this section.

UNLOADING (PREPARED BUT NOT FIRED)

B-38. See Table B-21 for more information about the procedures used to unload the M72AS subcaliber training launcher.

Table B-21. Procedures used to unload the M72AS subcaliber training launcher.

FIRER	ASSISTANT
WARNING **Keep the training launcher pointed downrange at all times.** (1) Return the trigger safety handle to the SAFE position, and wait for further instructions from the assistant.	**WARNING** **Keep your body out of the backblast area when inserting the M72AS subcaliber training launcher safety pin.** (1) Open the primer block cover to expose the training rocket primer block. (2) Carefully grasp the flash tube and pull rearward, removing the primer block and training rocket from the subcaliber training launcher. (3) Return the training rocket to the aluminum case, and return to the ASP at the end of the training day. (4) Instruct the firer to restore the training launcher to the carrying configuration.

MISFIRE PROCEDURES

B-39. See Table B-22 for more information about the procedures used to unload the M72AS subcaliber training launcher.

Table B-22. Procedures used to unload the M72AS subcaliber training launcher.

FIRER	ASSISTANT
WARNINGS **If a misfire occurs, wait 60 seconds.** **Do not remove the training launcher from the shoulder, and keep the training launcher pointed downrange at all times.** (1) If the training launcher fails to fire, release the trigger, press the trigger bar again, and hold. (2) If the training launcher fails to fire again, say "MISFIRE", and return the trigger arming handle to the SAFE position.	**WARNING** **Keep your body out of the backblast area.**

Table B-22. Procedures used to unload the M72AS subcaliber training launcher (continued).

FIRER	ASSISTANT
(3) Remove the launcher from the firing shoulder, keeping the training launcher pointed downrange. **CAUTION** Do not damage the front and rear sight when closing the M72AS training launcher. (4) Press the detent boot, and partially collapse the training launcher. *Note.* If the trigger safety handle will not return to the SAFE position after firing the training launcher, partially collapse the training launcher by depressing the detent boot, and then depress the trigger bar. An alternate method is to collapse the training launcher, pull forward on the trigger safety handle, and release it. (5) Hold the training launcher slightly away from the body, and grasp the rear sight cover firmly. (6) Extend the training launcher by moving your hands firmly in opposite directions. *Notes.* 1. Extending the training launcher too slowly can result in failure to cock the training launcher. 2. Be sure the training launcher snaps into the locked position. Check the tube lock by trying to push the tube together. The training launcher should not collapse if the training launcher is in the locked position. 3. If two attempts to lock the training launcher fail, close and set the training launcher aside for disposal by authorized ammunition personnel. (7) Place the training launcher on the firing shoulder. (8) Check the backblast area. (9) Arm, aim, and fire the training launcher. (10) If the training launcher fails to fire again, say "MISFIRE", and maintain the firing position for 60 seconds. (11) After 60 seconds have passed, return the trigger safety handle to the SAFE position, and wait for further instructions from the assistant.	(1) Open the primer block cover to expose the training rocket primer block. (2) Carefully grasp the flash tube and pull rearward to remove the primer block and training rocket from the subcaliber training launcher. (3) Instruct the firer to restore the training launcher to the carrying configuration. *Note.* If the transport pin cannot be reinserted, do not move the training launcher from the firing line. Notify authorized personnel for disposal of the training launcher and/or rocket.

MAINTAINING

B-40. Basic issue items are used for cleaning and maintaining shoulder-launched munition subcaliber training launchers. This is necessary to prevent carbon buildup and cracking, pitting, or damage to the subcaliber's rear end (carbon buildup may restrict firing pin movement).

Note. See the appropriate weapon TMs for authorized cleaning equipment and use.

SECTION III. M136 AT4 SIGHT ENGAGEMENT TRAINER

B-41. The M136 AT4 sight engagement trainer is used to train Soldiers to obtain a proper sight picture with the M136 AT4.

Notes. 1. The local TSC can provide GTA 07-02-005.

2. M136-series launchers and the M141 BDM have the same type front and rear sight. The M136 AT4 sight engagement trainer (GTA 07-02-005) can be used to train Soldiers on proper sight alignment for both launchers.

COMPONENTS

B-42. The M136 AT4 sight engagement trainer (GTA 07-02-005) consists of two parts:
- A target silhouette sheet.
- A front sight template.

B-43. The target silhouette sheet is a 8 1/2- by 12-inch piece of hard white plastic, and the front sight template is a 5 3/4- by 8 1/4 piece of transparent plastic (Figure B-11).

Note. The image depicted in Figure B-11 is a representation and not a true image.

USE

B-44. To use the sight engagement trainer—
- To obtain the correct sight picture for the target speed and range, place the front sight template on the target silhouette sheet as shown in Figure B-12.
- For each of the six targets, the reference number sheet shows the number that should appear in the circle on the front sight template.

Note. The reference number sheet is also used to determine the correct answer for any unit-developed test for determining the correct sight picture with the M136 AT4.

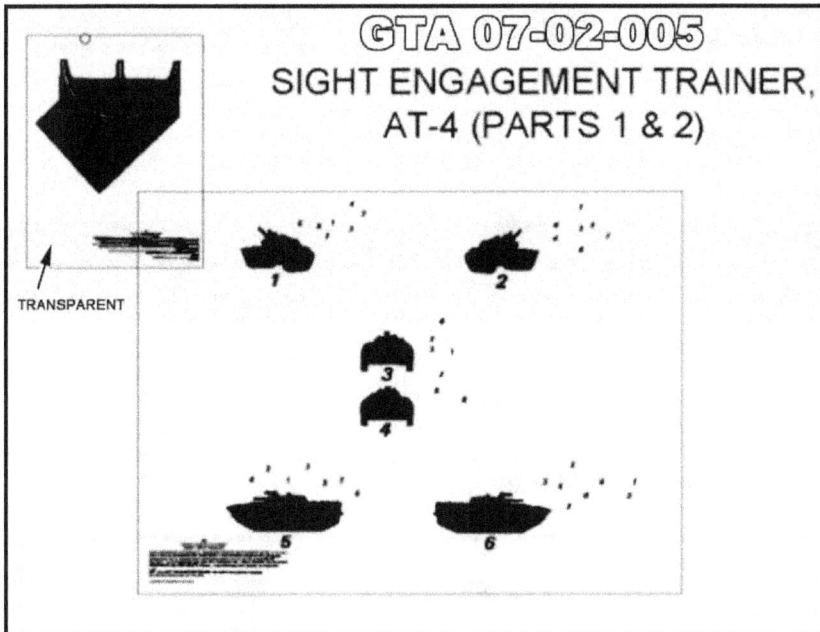

Figure B-11. Using GTA 7-02-005 with target silhouette sheet and front sight template.

Figure B-12. Using GTA 7-02-005 with target silhouette sheet and front sight template.

SECTION IV. ENGAGEMENT SKILLS TRAINER 2000

B-45. Unlike the FHT, the EST 2000 gives the Soldier the effects, conditions, and downrange feedback necessary for honing shoulder-launched marksmanship skills. EST 2000 replicates a shoulder-launched rocket's trajectory and provides an AAR of the firer's hit and miss in proximity of the target. With proper training and oversight by the instructor/trainer, a Soldier with poor marksmanship skills can improve with the help of the EST 2000.

SECTION V. MULTIPLE INTEGRATED LASER ENGAGEMENT SYSTEM

B-46. MILES simulators emulate the weight and operation of the weapon being simulated. Two shoulder-launched munitions are replicated in MILES: the M136 AT4 (Figure B-13) and the rocket-propelled grenade (RPG) launcher (Figure B-14). The M136 AT4 MILES simulator is used by the blue force (BLUFOR) and is the main shoulder-launched munition used for MILES force-on-force training; the RPG is the main shoulder-launched munition used by the opposing force (OPFOR).

> *Note.* For more information on the use and maintenance of the M136 AT4 and RPG launcher MILES simulators, see your post MILES facility or TSC.

WARNINGS

Do not look directly at the laser beam or the laser emitter through optics such as binoculars, telescopes, or periscopes at ranges of less than 75 meters (246 feet).

Do not look directly at the laser emitter at close range (less than 10 meters [32.8 feet]). Increasing the distance between the eye and the laser will reduce the risk of injury.

CAUTION
Do not connect or disconnect any cable while power is applied.

M136 AT4 SIMULATOR

B-47. Soldiers use the M136 AT4 MILES simulator (Figure B-13) to train tactics and operational familiarity. As such, it contains similar sights, triggers, safeties, and shouldering and holding elements.

> *Note.* Currently, there are no MILES simulators to replicate the M141 BDM or M72-series shoulder-launched munitions. The M136 AT4 MILES simulator can be used to simulate these shoulder-launched munitions during force-on-force training exercises.

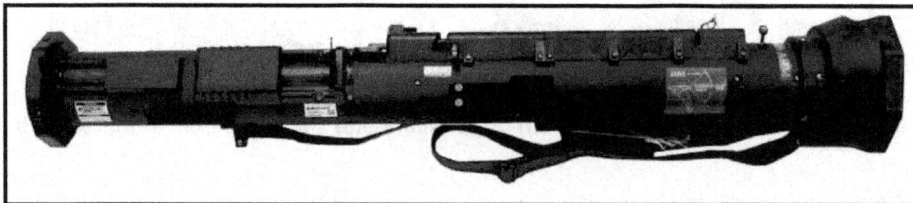

Figure B-13. M136 AT4 Multiple Integrated Laser Engagement System simulator.

ROCKET-PROPELLED GRENADE LAUNCHER SIMULATOR

B-48. Soldiers use the RPG MILES simulator (Figure B-14) to train tactics and operational familiarity, and provide more realistic OPFOR training. As such, it contains similar sights, triggers, safeties, and shouldering and holding elements.

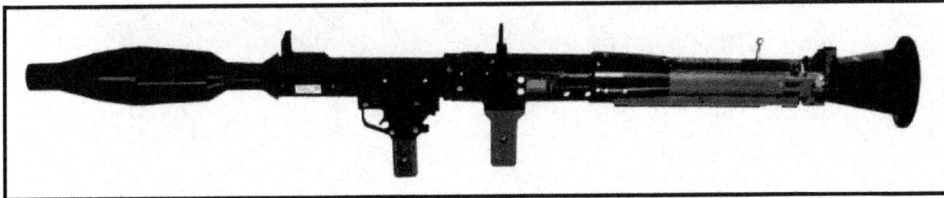

**Figure B-14. Rocket-propelled grenade launcher
Multiple Integrated Laser Engagement System simulator.**

PYROTECHNIC CUEING

B-49. In addition to looking like the actual munition and having the same triggering and safety elements, shoulder-launched munitions MILES simulators can provide a firing cue similar to the actual munition by inserting a standard MILES M22 cartridge (Figure B-15) prior to each firing. The system can be set to require the operator to insert an unexploded pyrotechnic in the MILES simulator or the simulated shoulder-launched munition will not fire.

WARNING

Pyrotechnic cueing devices can be dangerous; Soldiers should use them with caution. When using a M22 antitank weapon effects signature simulation (ATWESS) cartridge—

Treat a pyrotechnic-loaded shoulder-launched munitions MILES simulator as you would any loaded and armed munition.

DO NOT drop the simulator when a pyrotechnic is loaded and armed. A strong jolt may set off the ATWESS cartridge.

NEVER stand behind the simulator when arming or loading an ATWESS cartridge. DO NOT load a simulator unless you are preparing to fire.

Always assume that the simulator is armed and take appropriate safety measures.

The shoulder-launched munitions MILES simulator should not be armed until just prior to acquisition of the target.

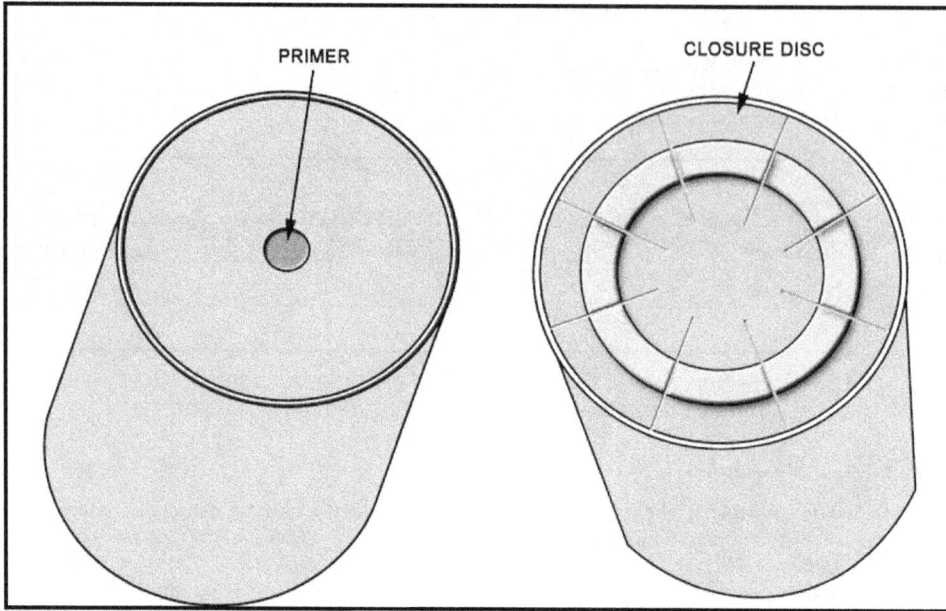

Figure B-15. M22 Multiple Integrated Laser Engagement System cartridge.

SURFACE DANGER ZONES

B-50. Pyrotechnic use with shoulder-launched munitions MILES simulators requires that SDZs be enforced to ensure a reasonable amount of safety (Figure B-16).

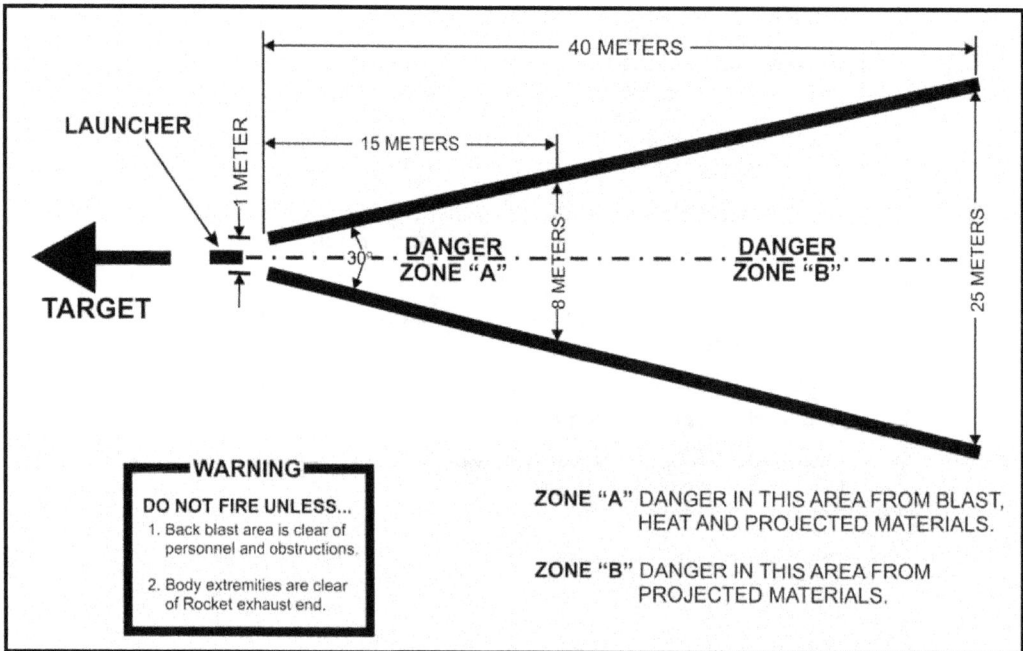

Figure B-16. M22 Multiple Integrated Laser Engagement System cartridge surface danger zones.

SAFETIES

B-51. To reduce operational risk, the M136 AT4 (Figure B-17) and the RPG launcher (Figure B-18) MILES simulators have two safety features:

- A pull-to-arm shaft.
- An ATWESS door.

Pull-to-Arm Shaft

B-52. The pyrotechnic will not fire unless the pull-to-arm shaft (1) is pulled up.

Antitank Weapon Effects Signature Simulation Door

B-53. When the ATWESS door (2) is opened, the ATWESS safety is automatically engaged.

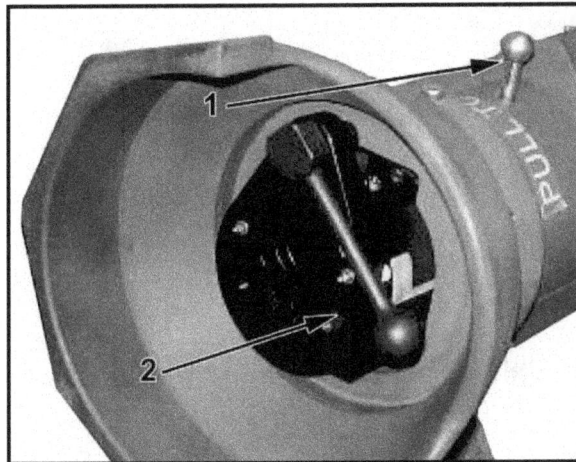

Figure B-17. M136 AT4 Multiple Integrated Laser Engagement System simulator safeties.

**Figure B-18. Rocket-propelled grenade launcher
Multiple Integrated Laser Engagement System simulator safeties.**

This page intentionally left blank.

Appendix C

NIGHT VISION DEVICES

Because shoulder-launched munitions are discarded after firing, the munition has no dedicated sight systems other than those permanently attached to the launcher. When conducting engagements in limited visibility conditions, Soldiers should use the various NVDs available. This appendix covers attaching and aligning NVDs to shoulder-launched munitions.

Notes.	1.	This appendix does not cover NVD operation. For more information about NVD operation, refer to the appropriate TMs.
	2.	If the NVD has been aligned correctly, there is generally no need to confirm using live ammunition.
	3.	Shoulder-launched munitions cannot be conventionally boresighted; therefore, the aided vision device must be aligned to the shoulder-launched munition.
	4.	Chapter 1 provides a list of NVDs that can be used with shoulder-launched munitions during limited visibility conditions.

MOUNTING NIGHT VISION DEVICES

C-1. The M141 BDM (Figure C-1) and the M136A1 AT4CS (Figure C-2) come equipped with a MIL-STD-1913 rail mount. The M141 BDM has only a side rail mount, but the M136A1 AT4CS has side and front rail mounts. This enables the M136A1 AT4CS to have a NVS and a laser aiming light mounted simultaneously; whereas, only one can be used on the M141 BDM at any given time.

C-2. M136 AT4s (Figure C-3) and M72-series shoulder-launched munitions (Figure C-4) require an attachable mounting bracket for NVD use. This mount accepts only one NVD at a time.

SIDE VIEW WITH AIMING LIGHT IN PLACE

RIFLE SIGHTS

DATA PLATE

MOUNTING
SCREWS/
RAIL

REAR VIEW

OPTICAL NIGHT
SIGHT

MOUNTING
RAIL

RIFLE
SIGHT

MOUNTING
SCREW

LAUNCHER

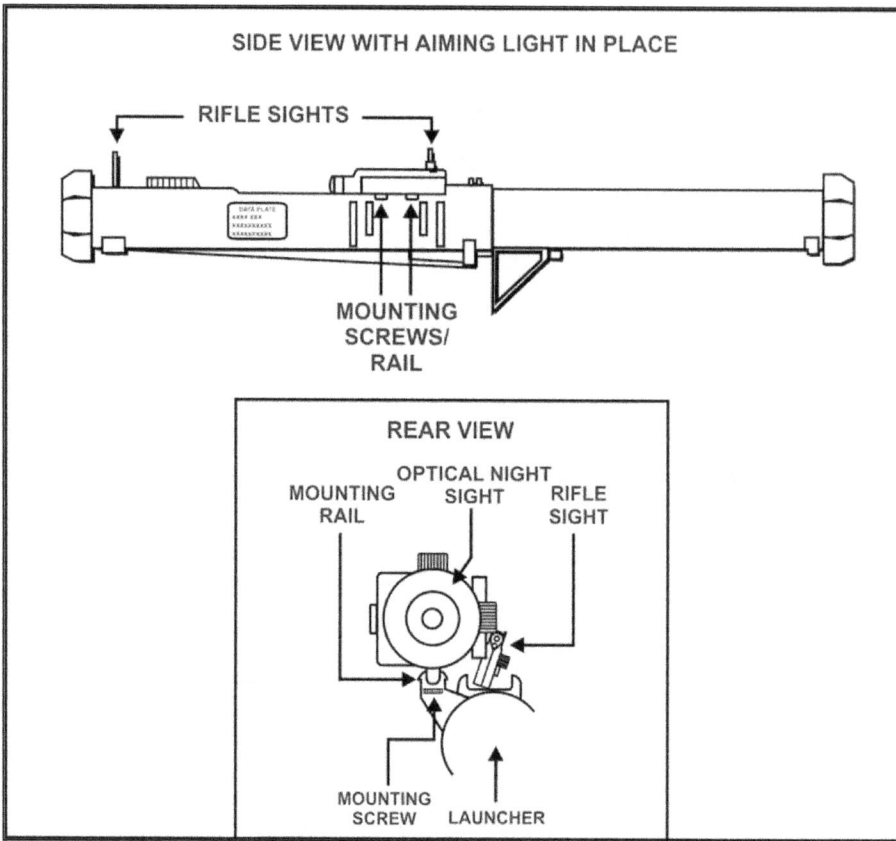

Figure C-1. M141 bunker defeat munition MIL-STD-1913 rail mount.

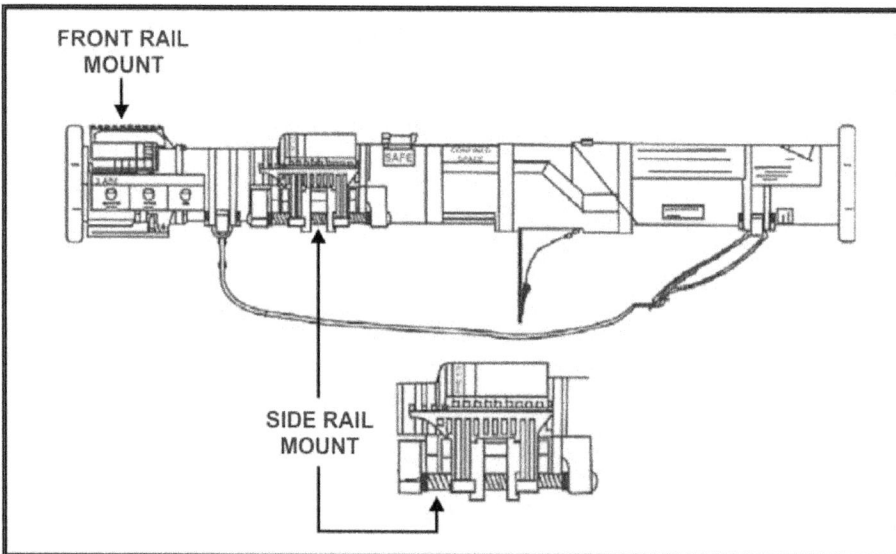

FRONT RAIL
MOUNT

SAFE

SIDE RAIL
MOUNT

Figure C-2. M136 AT4 MIL-STD-1913 rail mounts.

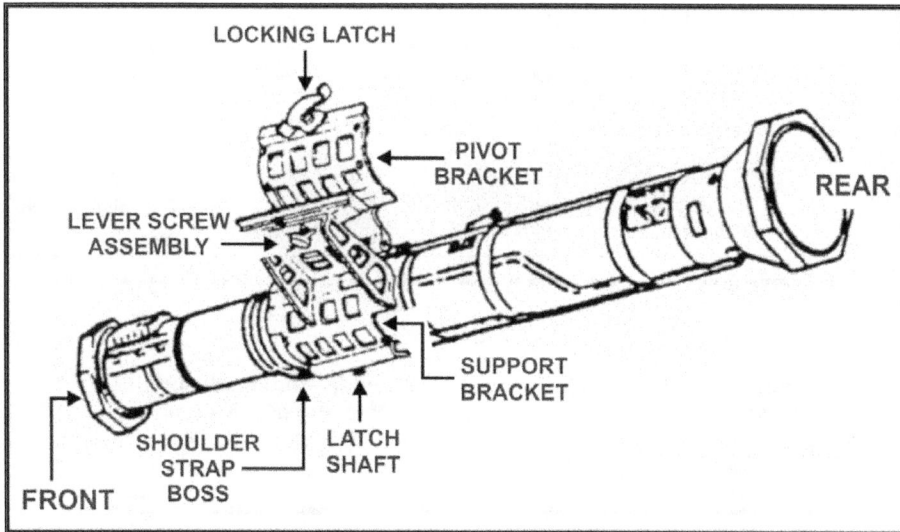

Figure C-3. M136 AT4 night vision device mounting bracket assembly.

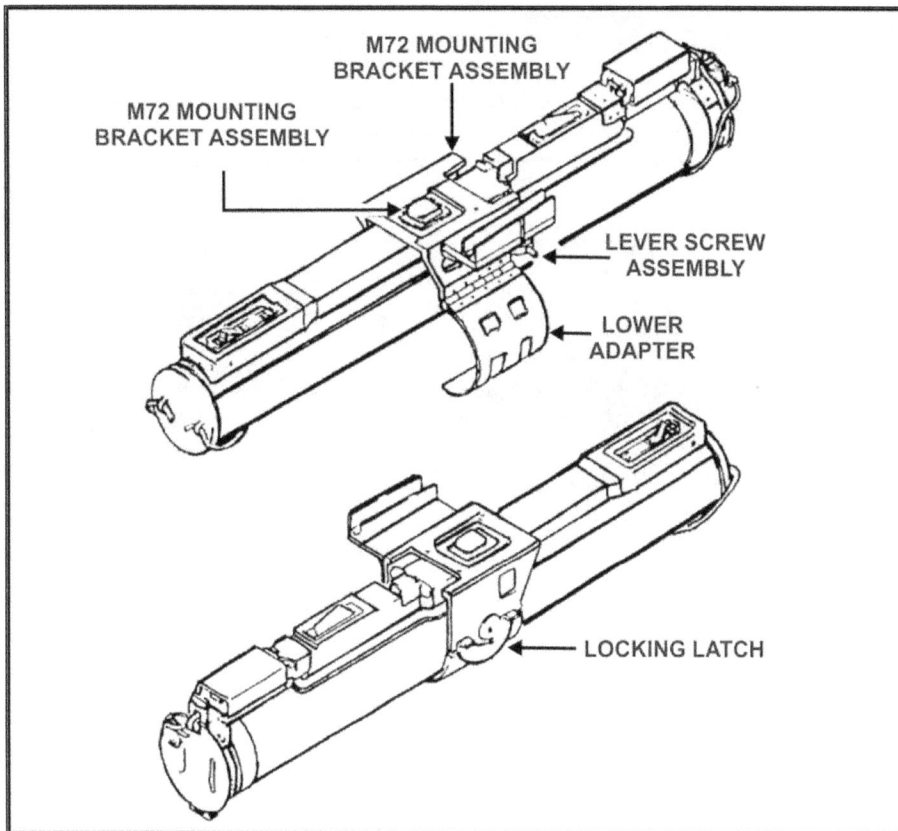

Figure C-4. M72-series night vision device mounting bracket assembly.

ALIGNMENT PROCEDURES

Notes.   1.   During NVD alignment procedures, an assistant qualified on shoulder-launched munitions should aid the firer by following his directions and providing assistance as he mounts the NVD and makes adjustments.

2.   The AN/PAS-13E LWTS (Figures C-5 and C-6) and the AN/PEQ-15 advanced target pointer/illuminator/aiming light (Figure C-7) are used to show shoulder-launched munition and NVD alignment procedures.

WARNING

When aligning any sight to a tactical shoulder-launched munition, be extremely careful to prevent accidental firing of the munition. Never cock the round or open the M141 BDM firing mechanism cover during this procedure.

C-3.   Alignment procedures for NVSs and laser aiming lights/illuminators can be performed during the day; however, you will need NVGs when aligning the laser aiming light. To successfully align shoulder-launched munitions to NVDs, the firer must—

- Select a firing platform that will support and stabilize the launcher during the alignment procedures.
- Have a clear field of view of a fixed, natural or manmade object at a known distance from the launcher. The minimum desired distance is 200 meters.

Figure C-5. AN/PAS-13E light weapon thermal sight.

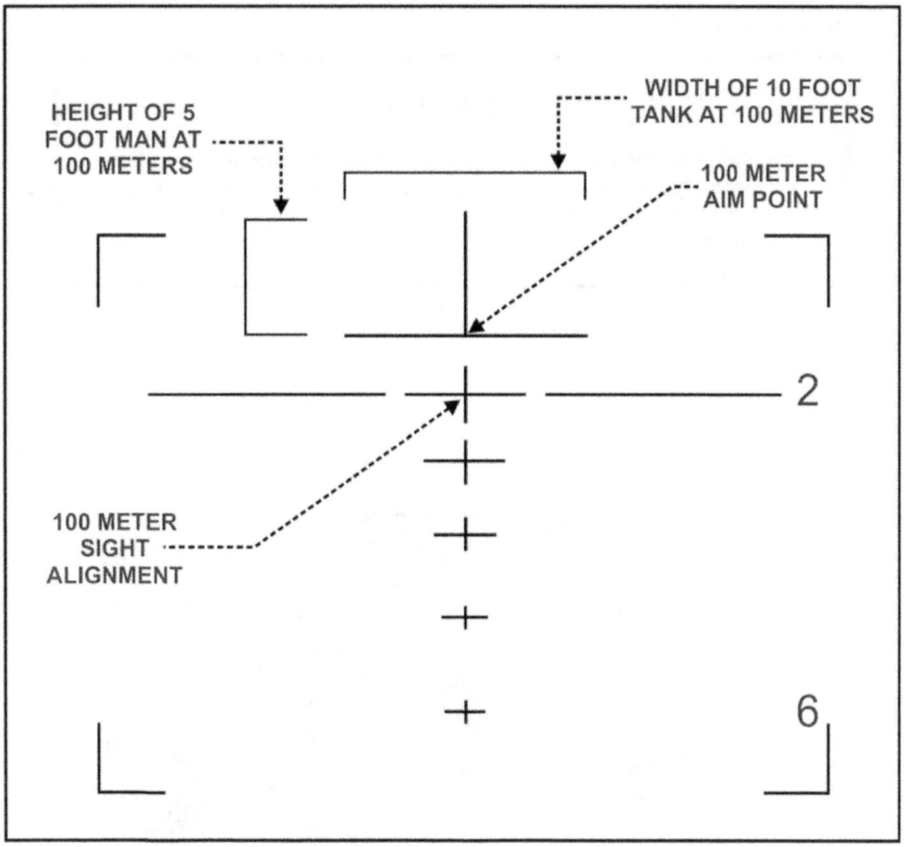
HEIGHT OF 5 FOOT MAN AT 100 METERS
WIDTH OF 10 FOOT TANK AT 100 METERS
100 METER AIM POINT
2
100 METER SIGHT ALIGNMENT
6

Figure C-6. AN/PAS-13E light weapon thermal sight reticle.

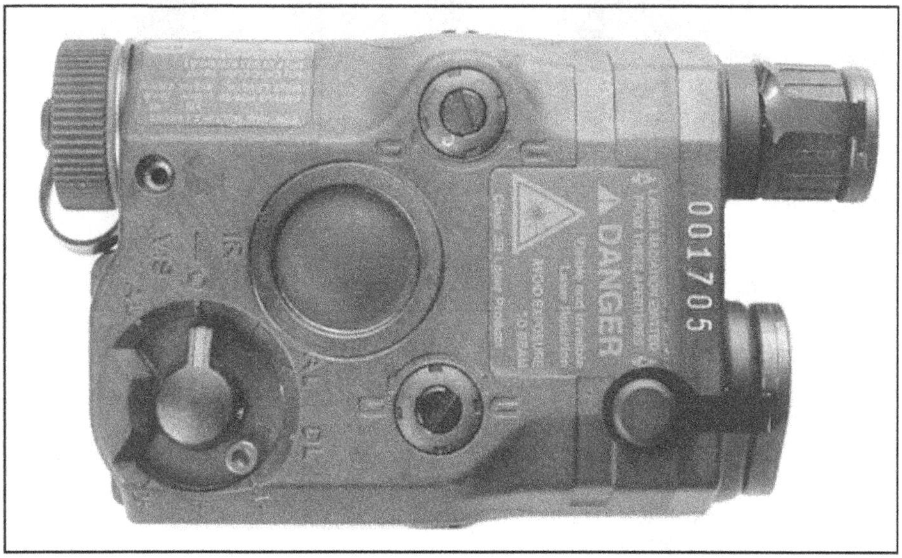
DANGER
001705

Figure C-7. AN/PEQ-15 advanced target pointer/illuminator/aiming light.

M141 BUNKER DEFEAT MUNITION

WARNING

When aligning any sight to a tactical shoulder-launched munition, be extremely careful to prevent accidental firing of the munition. Never open the M141 BDM firing mechanism cover during this procedure.

Night Vision Sight

C-4. To align a NVS to the M141 BDM—
 (1) Remove the protective cover from the launcher's side mounting rail.
 (2) Place the launcher on a stable platform.
 (3) While holding the launcher with your right hand, place the NVS on the mounting rail, and adjust for eye relief (Figure C-8).

Figure C-8. Finding night vision sight eye relief on the M141 bunker defeat munition.

 (4) Once eye relief has been determined, direct your assistant to secure the sight to the launcher rail.

Note. Tighten the sight limiter knob until you hear two audible clicks.

 (5) Turn the sight ON.

Note. See the appropriate TM for more information about system operation.

 (6) Adjust the launcher's position until the launcher's fixed sights are properly aimed at the selected object.
 (7) Adjust the alternate sight reticle until the aimpoint coincides with the launcher's fixed sight picture on the selected object.

Notes. 1. Be careful not to move the launcher during the alignment process. Verify that the launcher's fixed sights are still properly aligned with the selected object.

2. The AN/PVS-4 reticle will not be level when aligned to the M141 BDM. Carefully note the reticle angle—you will have to hold the launcher at the same angle when firing at night so that your firing will be accurate.

Laser Aiming Light/Illuminator

> **DANGER**
> DO NOT SHINE LASERS INTO YOUR EYES OR THE EYES OF OTHERS. THIS CAN CAUSE PERMANENT EYE INJURY.

Notes. 1. Laser aiming lights/illuminators project an IR laser beam that cannot be seen with the eye, but can be seen with NVSs/NVGs.

2. Conduct/verify sight alignment of laser aiming lights/illuminators after sundown. The laser beam is much easier to see during late evening hours.

C-5. To align the laser aiming light/illuminator to the M141 BDM—
(1) Place the laser aiming light/illuminator in the slot of the side mounting rail or engage the rail grabber.
(2) Adjust the laser aiming light/illuminator until one of the mounting screws engages the threaded hole, or place the rail grabber mount over the M141 BDM's alternate sight rail (Figure C-9).

Note. Tighten the sight limiter knob until you hear two audible clicks.

Figure C-9. Secure the laser aiming light/illuminator mounting screw or rail grabber.

(3) Turn the aiming light ON.

Note. See the appropriate TM for more information about system operation.

(4) Adjust the launcher's position until the launcher's fixed sights are properly aimed at the selected object.

(5) Adjust the laser aiming light/illuminator until the laser aimpoint coincides with the launcher's fixed sight picture on the selected object.

Note. Be careful not to move the launcher during the alignment process. Verify that the launcher's fixed sights are still properly aligned with the selected object.

M136 AT4

WARNING

When aligning any sight to a tactical shoulder-launched munition, be extremely careful to prevent accidental firing of the munition. Never cock the round during this procedure.

Night Vision Sight

C-6. To align a NVS to the M136 AT4—
(1) Attach the NVD mounting bracket to the launcher (Figure C-10).
- Cradle the M136 AT4 in your left arm.
- Position the support bracket with the mounting rail on the left side and the marking FRONT over the rear sight cover.
- With the pivot bracket spread open, place the support bracket against the base of the rear sight housing and the bottom on the shoulder strap boss.
- Swing the pivot bracket around the M136 AT4, and secure it by rotating the locking latch clockwise to engage the latch shaft.
- Place the bracket adapter in the groove of the mounting rail so that the threaded screw hole in the base of the adapter is aligned with the lever screw assembly. Tighten the lever screw.
(2) Place the launcher on the shoulder.
(3) While holding the launcher with your right hand, place the NVS on the mounting rail, and adjust for eye relief (Figure C-11).
(4) Once eye relief has been determined, direct your assistant to secure the sight to the launcher rail.

Note. Tighten the sight limiter knob until you hear two audible clicks.

(5) Take the launcher off the shoulder.
(6) Turn the sight ON.

Note. See the appropriate TM for more information about system operation.

(7) Adjust the launcher's position until the launcher's fixed sights are properly aimed at the selected object.
(8) Adjust the alternate sight reticle until the aimpoint coincides with the launcher's fixed sight picture on the selected object.

Note. Be careful not to move the launcher during the sight alignment process. Verify that the launcher's fixed sights are still properly aligned with the selected object.

Figure C-10. M136 AT4 with night vision device mounting bracket attached.

Figure C-11. Finding night vision sight eye relief on the M136 AT4.

Laser Aiming Light/Illuminator

> **DANGER**
>
> DO NOT SHINE LASERS INTO YOUR EYES OR THE EYES OF
> OTHERS. THIS CAN CAUSE PERMANENT EYE INJURY.

Notes.   1.   Laser aiming lights/illuminators project an IR laser beam that cannot be seen with the eye, but can be seen with NVSs/NVGs.

2.   Conduct/verify sight alignment of laser aiming lights/illuminators after sundown. The laser beam is much easier to see during late evening hours.

C-7.   To align the laser aiming light/illuminator to the M136 AT4—
(1)   Place the laser aiming light/illuminator in slot of the launcher front mounting rail or engage the rail grabber.

Note. The lever screw assembly must be located in the rear threaded screw hole when mounting the laser aiming light/illuminator. Lever screw assembly may require relocation from the front to the rear threaded screw hole.

(2)   Adjust the laser aiming light/illuminator until one of the mounting screws engages the threaded hole, or place the rail grabber mount over the alternate sight rail (Figure C-12).

Note. Tighten the sight limiter knob until you hear two audible clicks.

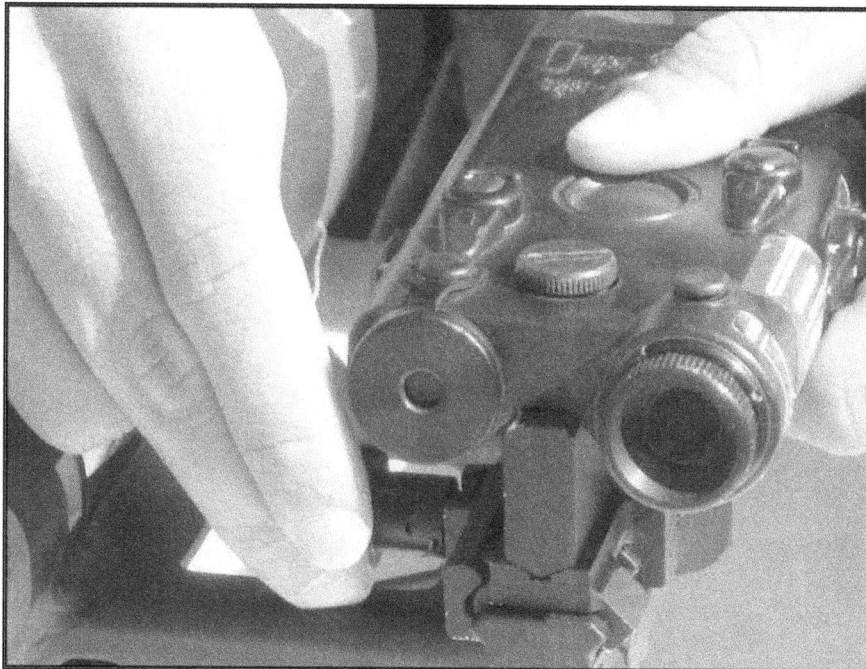

Figure C-12. Securing the laser aiming light/illuminator to the M136 AT4.

(3) Turn the sight ON.

Note. See the appropriate TM for more information about system operation.

(4) Adjust the launcher's position until the launcher's fixed sights are properly aimed at the selected object.
(5) Adjust the laser aiming light/illuminator until the laser aimpoint coincides with the launcher's fixed sight picture on the selected object, or use an assistant to align the laser aiming light/illuminator to the target.

Note. Be careful not to move the launcher during the alignment process. Verify that the launcher's fixed sights are still properly aligned with the selected object.

M136A1 AT4 CONFINED SPACE

WARNING

When aligning any sight to a tactical shoulder-launched munition, be extremely careful to prevent accidental firing of the munition. Never cock the round during this procedure.

Night Vision Sight

C-8. To align a NVS to the M136A1 AT4CS—
(1) Lift and lock the folding side rail mount.

Note. See TM 9-1315-255-13 for more information about operating the side mounting bracket.

(2) Place the launcher on the shoulder.
(3) While holding the launcher with your right hand, place the NVS on the mounting rail and adjust for eye relief (Figure C-13).
(4) Once eye relief has been determined, direct your assistant to secure the sight to the launcher rail.

Note. Tighten the sight limiter knob until you hear two audible clicks.

(5) Take the launcher off the shoulder.
(6) Turn the sight ON.

Note. See the appropriate TM for more information about system operation.

(7) Adjust the launcher's position until the launcher's fixed sights are properly aimed at the selected object.
(8) Adjust the alternate sight reticle until the aimpoint coincides with the launcher's fixed sight picture on the selected object.

Note. Be careful not to move the launcher during the alignment process. Verify that the launcher's fixed sights are still properly aligned with the selected object.

Figure C-13. Finding night vision sight eye relief on the M136A1 AT4 confined space.

Laser Aiming Light/Illuminator

> # DANGER
> DO NOT SHINE LASERS INTO YOUR EYES OR THE EYES OF OTHERS. THIS CAN CAUSE PERMANENT EYE INJURY.

Notes. 1. Laser aiming lights/illuminators project an IR laser beam that cannot be seen with the eye, but can be seen with passive NVSs/NVGs.

2. Conduct/verify sight alignment of laser aiming lights/illuminators after sundown. The laser beam is much easier to see during late evening hours.

C-9. To align the laser aiming light/illuminator to the M136A1 AT4CS—
(1) Place the laser aiming light/illuminator in the slot of the launcher front mounting rail or engage the rail grabber (Figure C-14).
(2) Adjust the laser aiming light/illuminator until one of the mounting screws engages the threaded hole, or place the rail grabber mount over the mounting rail. Securely hand-tighten the mounting screw or rail grabber.

Note. Tighten the sight limiter knob until you hear two audible clicks.

(3) Turn the sight ON.

Note. See the appropriate TM for more information about system operation.

(4) Adjust the launcher's position until the launcher's fixed sights are properly aimed at the selected object. Use an assistant to align the laser aiming light/illuminator to the target.

(5) Adjust the laser aiming light/illuminator until the laser aimpoint coincides with the launcher's fixed sight picture on the selected object.

Note. Be careful not to move the launcher during the alignment process. Verify that the launcher's fixed sights are still properly aligned with the selected object.

Figure C-14. Securing the laser aiming light/illuminator to the M136A1 AT4 confined space.

M72-Series Shoulder-Launched Munitions

WARNING

When aligning any sight to a tactical shoulder-launched munition, be extremely careful to prevent accidental firing of the munition. Never cock the round during this procedure.

Night Vision Sight

C-10. To align a NVS to M72-series launchers—

(1) Attach the NVS mounting bracket to the launcher (Figure C-15).

- Place the bracket assembly on top of the launcher so that the square cutout in the top of the bracket fits over the extension release button.
- Swing the lower adapter section up and under the rocket launcher, and secure it by turning the locking latch clockwise to fully engage the latch shoulder screw.
- Place the sight in the groove on the side bracket, and align the threaded screw hole in the base of the sight with the lever screw assembly.

Note. Tighten the sight limiter knob until you hear two audible clicks.

- Tighten the lever screw assembly firmly.

(2) Extend the launcher.

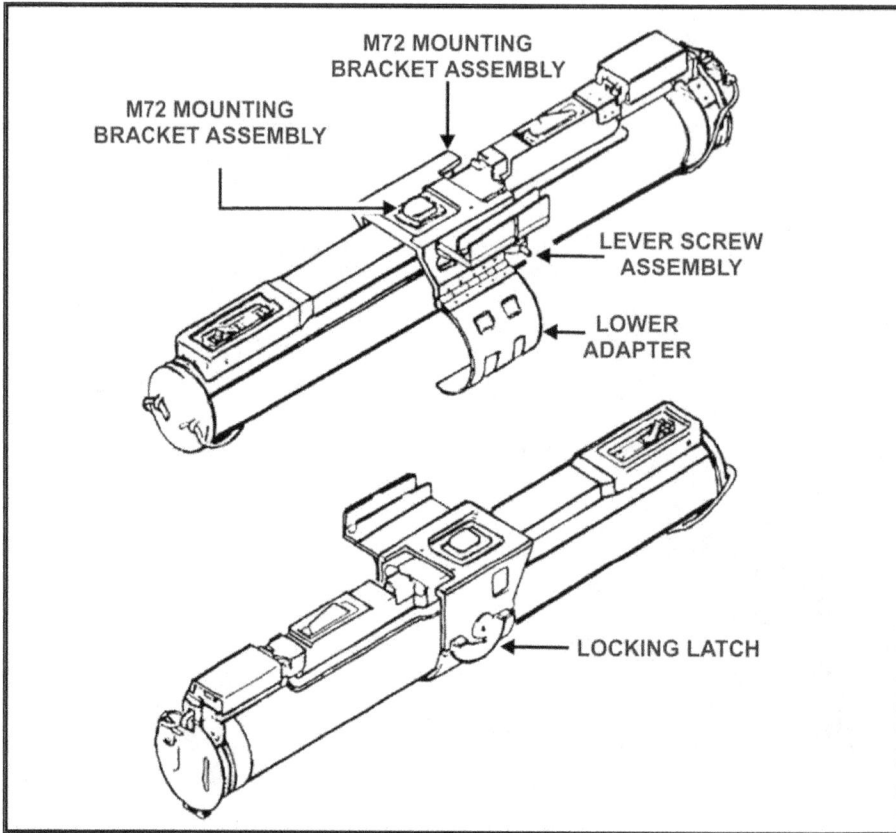

Figure C-15. M72-series attachable night vision device mounting bracket.

(3) Place the launcher on the shoulder.

(4) While holding the launcher with your right hand, place the NVS on the mounting rail and adjust for eye relief.

(5) Once eye relief has been determined, direct your assistant to secure the sight to the mounting bracket. Tighten the sight limiter knob until you hear two audible clicks.

(6) Take the launcher off the shoulder.

(7) Turn the sight ON.

Note. See the appropriate TM for more information about system operation.

(8) Adjust the launcher's position until the launcher's fixed sights are properly aimed at the selected object.

(9) Adjust the alternate sight reticle until the aimpoint coincides with the launcher's fixed sight picture on the selected object.

Note. Be careful not to move the launcher during the sight alignment process. Verify that the launcher's fixed sights are still properly aligned with the selected object.

Laser Aiming Light/Illuminator

> **DANGER**
> DO NOT SHINE LASERS INTO YOUR EYES OR THE EYES OF
> OTHERS. THIS CAN CAUSE PERMANENT EYE INJURY.

Notes. 1. Laser aiming lights/illuminators project an IR laser beam that cannot be seen with the eye, but can be seen with NVSs/NVGs.

2. Conduct/verify sight alignment of laser aiming lights/illuminators after sundown. The laser beam is much easier to see during late evening hours.

C-11. To align the laser aiming light/illuminator to M72-series launchers —
(1) Place the laser aiming light/illuminator on the side mounting rail or front rail on the M72A6/A7.
(2) Adjust the laser aiming light/illuminator until one of the mounting screws engages the threaded hole, or place the rail grabber mount over the alternate sight rail.
(3) Securely hand-tighten the mounting screw or rail grabber.

Note. Tighten the sight limiter knob until you hear two audible clicks.

(4) Turn the aiming light ON.

Note. See the appropriate TM for more information about system operation.

(5) Adjust the launcher's position until the launcher's fixed sights are properly aimed at the selected object.
(6) Adjust the laser aiming light/illuminator until the laser aimpoint coincides with the launcher's fixed sight picture on the selected object.

Note. Be careful not to move the launcher during the alignment process. Verify that the launcher's fixed sights are still properly aligned with the selected object.

This page intentionally left blank.

GLOSSARY

A

AR	Army regulation
ARNG	Army National Guard
ARNGUS	Army National Guard of the United States
ASP	ammunition supply point
AT4CS	AT4 confined space
ATWESS	antitank weapon effects signature simulation

B

BDM	bunker defeat munition

C

C	Celsius
CAE	combat arms earplug
CALFEX	combined arms live-fire exercise
CBRN	chemical, biological, radiological, nuclear
cm	centimeter
CRM	composite risk management

D

DA	Department of the Army
DA PAM	Department of the Army pamphlet
DAP	decontaminating apparatus, portable
DS2	decontamination solution 2

E

EFP	explosively formed penetrator
EOD	explosive ordnance disposal
EST 2000	Engagement Skills Trainer 2000
ETLBV	enhanced tactical load-bearing vest

F

F	Fahrenheit
FET	field-expedient trainer
FHT	field handling trainer
FM	field manual
FTX	field training exercise

G

GS TM	general subject technical manual
GTA	graphic training aid

H

HE	high-explosive
HEAT	high-explosive antitank
HEDP	high-explosive, dual purpose
HWTS	heavy weapon thermal sight

I

IEDK	individual equipment decontamination kit
IET	initial entry training
IR	infrared

K

km	kilometer

L

LCE	load-carrying equipment
LFX	live-fire exercise
LWTS	light weapon thermal sight

M

MCoE	Maneuver Center of Excellence
METL	mission-essential task list
METT-TC	mission, enemy, terrain, troops, time available, and civil considerations
MILES	Multiple Integrated Laser Engagement System
mm	millimeter
MOPP	mission-oriented protective posture
MPRC	multipurpose range complex

N

NCO	noncommissioned officer
NCOIC	noncommissioned officer in charge
NSN	national stock number
NVD	night vision device
NVGs	night vision goggles
NVS	night vision sight

O

OIC	officer in charge
OPFOR	opposing force

P

PDDE	power-driven decontamination equipment
POV	privately-owned vehicle

R

RPG	rocket-propelled grenade
RSO	range safety officer

S

SDK	skin decontaminating kit
SDZ	surface danger zone
SMCT	Soldier's manual of common tasks
SOP	standing operating procedure
STB	supertropical bleach
STP	Soldier training publication
STX	situational training exercise

T

TADSS	training aids, devices, simulators, and simulations
TC	training circular
TM	technical manual
TP-T	training practice-tracer
TRADOC	Training and Doctrine Command
TSC	training support center
TWS	thermal weapon sight

U

USAR	United States Army Reserve

REFERENCES

SOURCES USED

These are the sources quoted or paraphrased in this manual:

AR 350-1, *Army Training and Leader Development*, 18 December 2009.

DA PAM 350-38, *Standards in Training*, 13 May 2009.

DA PAM 385-63, *Range Safety,* 4 August 2009.

FM 3-11.5, Multiservice Tactics, Techniques, and Procedures for Chemical, Biological, Radiological, and Nuclear Decontamination. 04 April 2006.

FM 3-22.9, Rifle Marksmanship M16-/M4-Series Weapons. 12 August 2008.

FM 3-25.26, Map Reading and Land Navigation. 18 January 2005.

FM 3-34.214, Explosives and Demolitions. 11 July 2007.

FM 5-19, Composite Risk Management. 21 August 2006.

FM 7-0, Training for Full Spectrum Operations. 12 December 2008.

STP 21-1-SMCT, *Soldier's Manual of Common Tasks, Skill level 1,* 18 June 2009.

STP 21-24-SMCT, *Soldier's Manual of Common Tasks (SMCT) Warrior Leader, Skill Level 2, 3, and 4,* 9 September 2008.

TB 9-1340-230-13, Operator's and Field Information for Rocket, High Explosive, 66 Millimeter: Light Anti-Armor Weapon (LAW), HEAT, M72A7. 31 December 2007.

TC 25-8, Training Ranges. 20 May 2010.

TM 9-1055-886-12&P, Operator and Unit Maintenance Manual (Including Repair Parts and Special Tools List) for M287 9-mm Tracer Bullet Training Device (NSN 1055-01-207-2684). 25 October 1989.

TM 9-1315-255-13, Operator and Field Maintenance Manual for Launcher and Cartridge, 84 Millimeter: AT4 Confined Space and Reduced Sensitivity (AT4CS-RS), M136A1 (NSN: 1315-01-508-8521). 19 June 2009.

TM 9-1315-886-12, Operator's and Unit Maintenance Manual for Launcher and Cartridge, 84 Millimeter: M136 (AT4). 15 May 1990.

TM 9-1340-214-10, Operator's Manual for 66mm Light Antitank Weapon (LAW) System M72A1, and M72A2 with Coupler, M72A3 and Practice Rocket Launcher M190 with M73 Practice Rocket. 31 May 1991.

TM 9-1340-228-10, Operator's Manual For Rocket And Launcher 83 Millimeter: HEDP (SMAW-D), M141 (NSN 1340-01-443-5477). 01 September 2005.

TM 11-5855-312-10, Operator's Manual: Sight, Thermal AN/PAS-13B(V)2 (NSN 5855-01-464-3152); AN/PAS-13B(V)3 (5855-01-464-3151). 15 February 2005.

TM 11-5855-316-10, Operator's Manual: AN/PAS-13C(V)1 Sight, Thermal (NSN 5855-01-523-7707); AN/PAS-13C(V)2 Sight, Thermal (NSN 5855-01-523-7713); AN/PAS-13C(V)3 Sight, Thermal (NSN 5855-01-523-7715). 01 September 2010.

TM 11-5855-317-10, Operator's Manual for Sight, Thermal AN/PAS-13D(V)2 (NSN 5855-01-524-4313) (EIC: JH5); AN/PAS-13D(V)3 (NSN 5855-01-524-4314). 15 May 2009.

DOCUMENTS NEEDED

These documents must be available to the intended users of this publication.

DA Form 7676, *Day and Night Fire—M141 BDM (BDM Subcaliber Training Launcher)*.

DA Form 7677, *Day and Night Fire—M136 AT4 (M287 Subcaliber Training Launcher)*.

DA Form 7678, *Day and Night Fire—M72 (M72AS 21-mm Subcaliber Training Launcher)*.

DA Form 2028, *Recommended Changes to Publications and Blank Forms*.

GTA 7-02-005, *Sight Engagement Trainer*, M136 AT4 (Parts 1 and 2), 02 January 1990.

READINGS RECOMMENDED

These readings contain relevant supplemental information.

AR 190-11, *Physical Security of Arms, Ammunition, and Explosives*, 15 November 2006.

AR 385-63, *Range Safety*, 19 May 2003.

FM 3-06.11, *Combined Arms Operations in Urban Terrain*, 28 February 2002.

FM 3-21.71, *Mechanized Infantry Platoon and Squad (Bradley)*, 20 August 2002.

FM 3-21.8, *The Infantry Platoon and Squad*, 28 March 2007.

FM 20-3, *Camouflage, Concealment, and Decoys*, 30 August 1999.

FM 90-5, *Jungle Operations*, 16 August 1982.

MIL-STD-1913, *Military Standard: Dimensioning of Accessory Mounting Rail for Small Arms Weapons*, 3 February 1995.

Unit SOPs

INTERNET WEBSITES

Reimer Digital Library, https://atiam.train.army.mil/

Army Publishing Directorate, http://www.apd.army.mil/

Index

www.ingramcontent.com/pod-product-compliance
Lightning Source LLC
Chambersburg PA
CBHW081501200326
41518CB00015B/2342